The Mössbauer Effect
(Second Edition)

Online at: https://doi.org/10.1088/978-0-7503-6039-5

The Mössbauer Effect (Second Edition)

Richard A Dunlap
Department of Physics and Atmospheric Science, Dalhousie University, Halifax, Nova Scotia, Canada

IOP Publishing, Bristol, UK

ISBN 978-0-7503-6039-5 (ebook)
ISBN 978-0-7503-6037-1 (print)
ISBN 978-0-7503-6040-1 (myPrint)
ISBN 978-0-7503-6038-8 (mobi)

DOI 10.1088/978-0-7503-6039-5

Version: 20231201

IOP ebooks

British Library Cataloguing-in-Publication Data: A catalogue record for this book is available from the British Library.

Published by IOP Publishing, wholly owned by The Institute of Physics, London

IOP Publishing, No.2 The Distillery, Glassfields, Avon Street, Bristol, BS2 0GR, UK

US Office: IOP Publishing, Inc., 190 North Independence Mall West, Suite 601, Philadelphia, PA 19106, USA

Contents

Preface

The effect which now bears his name was discovered in 1958 by Rudolf Mössbauer at the Technical University of Munich. At first, this appeared to be a phenomenon related to nuclear energy levels that provided some information about excited state lifetimes and quantum properties. However, it soon became apparent that Mössbauer spectroscopy had applications in such diverse fields as general relativity, solid state physics, chemistry, materials science, biology, medical physics, archeology and art. It is the extreme sensitivity of the effect to the atomic environment around the probe atom as well as the ability to apply the technique to some interesting and important elements, most notably iron that are responsible for the Mössbauer effect's extensive use. The first edition of the present book reviewed the historical development of the Mössbauer effect, the experimental details, the basic physics of hyperfine interactions and some of the numerous applications of Mössbauer effect spectroscopy.

This second edition includes a new chapter on experimental methods and updated and greatly expanded material on applications of the Mössbauer effect. The latter includes a new chapter on modern techniques such as synchrotron radiation and implantation studies.

Acknowledgments

I am grateful to all the students and collaborators whom I have interacted with over the years in the field of Mössbauer spectroscopy. I would particularly like to thank the following: Dhiren Bahadur, Jeff Dahn, Tim Hatchard, Chris Hohenemser, Shacheenatha Jha, Derek Lawther, Bill Leiper, Jerry MacKay, Mark Obrovac, Bob O'Handley, Zbyszek Stadnik, Gerhard Stroink, and Srinivas Veeturi. For many years their contributions and encouragement formed an important component of my research involving the truly amazing experimental technique that forms the subject of the present book.

Author biography

Richard A Dunlap

Richard A Dunlap received a B.S. in Physics from Worcester Polytechnic Institute in 1974, an A.M. in Physics from Dartmouth College in 1976 and a PhD in Physics from Clark University in 1981. Since receiving his PhD, he has been on the Faculty at Dalhousie University. He was appointed Faculty of Science Killam Research Professor in Physics from 2001 to 2006 and served as Director of the Dalhousie University Institute for Research in Materials from 2009 to 2015. He currently holds an appointment as Research Professor in the Department of Physics and Atmospheric Science. Professor Dunlap has published more than 300 refereed research papers and his research interests have included, magnetic materials. amorphous alloys, critical phenomena, hydrogen storage, quasicrystals, superconductivity and materials for advanced batteries. Much of his work involves the application of Mössbauer effect spectroscopy to the investigation of solid state properties. He has been a member of the Mössbauer Century Club since 2005. Professor Dunlap's previous books include *Experimental Physics: Modern Methods* (Oxford 1988), *The Golden Ratio and Fibonacci Numbers* (World Scientific 1997), *An Introduction to the Physics of Nuclei and Particles* (Brooks/Cole 2004), *Sustainable Energy* (Cengage, 1st edn 2015, 2nd edn 2019), *Novel Microstructures for Solids* (IOP/Morgan & Claypool 2018), *Particle Physics* (IOP/Morgan & Claypool 2018), *The Mössbauer Effect* (IOP/Morgan & Claypool 2019), *Electrons in Solids: Contemporary Topics* (IOP/Morgan & Claypool 2019), *Energy from Nuclear Fusion* (IOP Publishing 2021), *Transportation Technologies for a Sustainable Future* (IOP Publishing 2023) and *Lasers and Their Application in the Cooling and Trapping of Atoms—2nd edn* (IOP Publishing 2023).

IOP Publishing

The Mössbauer Effect (Second Edition)

Richard A Dunlap

Chapter 1

The history of resonance fluorescence

1.1 Introduction

The phenomenon now known as the Mössbauer effect was first discovered by Rudolf Mössbauer (1929–2011) in 1958 while he was a doctoral student at the Technical University of Munich (Technische Universität München). When he first published the result of his discovery, he referred to the effect that he had observed as 'Kernresonanzfluoreszenz von Gammastrahlung mit verschwindenden Rückstoβenergieverlustes'. Literally translated this means 'Nuclear resonance fluorescence of gamma radiation with vanishing recoil energy loss' or (more succinctly) 'Recoil-free nuclear resonance fluorescence'. This terminology uniquely defines the effect that Mössbauer sometimes described as 'like throwing a stone out of a rowboat in the wintertime'. In the present chapter, we introduce the concept of resonance fluorescence, first with reference to atomic resonance fluorescence and then nuclear resonance fluorescence. The importance of the Heisenberg linewidth is also discussed.

1.2 Atomic resonance fluorescence

The earliest observation of a phenomenon that is based on atomic resonance fluorescence occurred in the early part of the 19th century. In 1802 the British scientist William Hyde Wollaston (1766–1828) was observing the optical spectrum of the sun and noticed the presence of dark lines in the continuum spectrum. A few years later, around 1814, Joseph von Fraunhofer (1787–1826) observed the same phenomenon in the optical spectra of certain gases in the laboratory. A continuum source, such as a flame, showed narrow absorption lines when passed through a region of cold gas. In his investigations Fraunhofer identified the characteristic spectral lines associated with various elements.

In the following years, it was discovered that a gas heated to a high temperature would emit electromagnetic radiation at specific wavelengths giving rise to a spectrum consisting of bright lines at well defined wavelengths on a dark background. In 1859 Gustav Kirchhoff (1824–87) proposed what is now known as

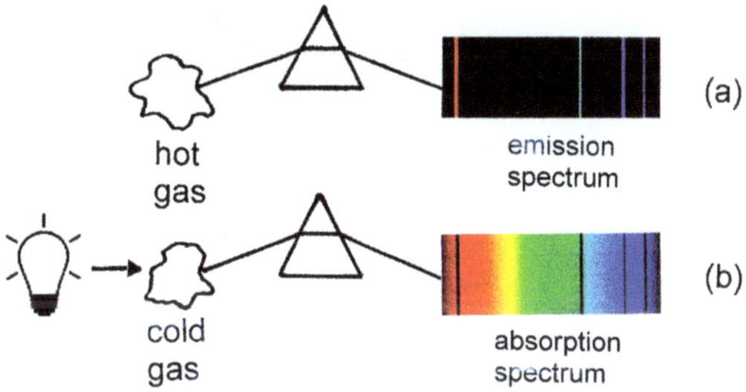

Figure 1.1. Observation of (a) an emission spectrum from a hot gas and (b) an absorption spectrum from a cold gas.

Kirchhoff's law (or sometimes Kirchhoff's law of spectroscopy, to distinguish it from the well known Kirchhoff's laws related to currents in electrical circuits). Kirchhoff's law states that an element which, as a cold gas, produces an absorption spectrum with dark lines at particular wavelengths will produce an emission spectrum when heated with emission lines at those same wavelengths. This behavior is illustrated in figure 1.1 where the properties of the resulting spectrum can be related to the details of the light source. Nearly a century after Wollaston's observation of solar spectral absorption lines, the relationship between the discrete spectra associated with specific elements was understood on the basis of the electronic structure of atoms.

Figure 1.2 shows a slightly different arrangement for obtaining optical spectra which provides considerable insight into the nature of atomic transitions. At point (a) in figure 1.2 an absorption spectrum is observed as in figure 1.1(b). However, an emission spectrum is observed at point (b) in figure 1.2. These two different spectra can be explained in the following way. Photons from the continuum source, which have just the right energy, will excite atomic transitions in the gas atoms. In this way, the energy of the photon will be absorbed, thus preferentially eliminating that particular wavelength from the continuum spectrum and giving rise to the dark lines in the spectrum observed in the forward direction from the source to the spectrum at point (a). The atoms which have excited electronic levels as a result of absorbing photons are unstable and will spontaneously decay back to their ground state. In this process, the atoms will release photons of the same energy (more or less) as the photons that had been absorbed. The question of how closely the absorbed and emitted photon energies are to one another should be considered. This point will be of fundamental importance in the discussion of nuclear resonance fluorescence below, but for now it will be suitable to merely view the photon energies as the same. The absorbed photons were travelling in a line from the source to point (a), however, the re-emitted photons are re-emitted isotopically, that is, uniformly in all spatial directions. These re-emitted photons give rise to the emission spectrum at point (b). They also contribute to the wavelengths of the absorption lines at point (a).

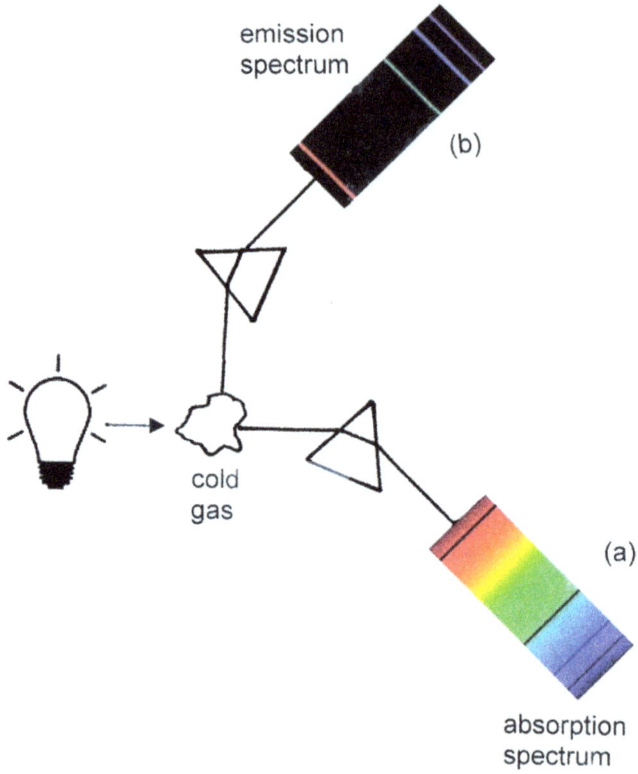

Figure 1.2. Geometry illustrating (a) absorption spectrum of continuum spectrum passing through cold gas and (b) spectrum of re-emitted photons, as described in the text.

However, as only a small fraction of the re-emitted photons will be travelling along the line from the source to point (a), then there is a net decrease caused by the absorption at these wavelengths in this direction, giving rise to the dark lines.

In 1904 Robert W Wood (1868–1955) elaborated on the concept of the emission spectrum shown at point (b) in figure 1.2 and described the phenomenon of atomic resonance fluorescence as illustrated in figure 1.3. In figure 1.3(a) a hot gas (the source) produces an emission spectrum characteristic of the element in the gas which is then observed as bright emission lines. In figure 1.3(b) a sample of cold gas of the same element (the absorber) is placed between the hot gas and the spectrometer. Since the atomic energy levels in the absorber are the same as they are in the source, the cold gas will preferentially absorb the photons produced by the source, thus eliminating (more or less) the bright emission lines in the resulting spectrum.

Figure 1.4 shows the atomic resonance fluorescence phenomenon from an atomic standpoint. The source atom undergoes a decay from the excited state to the ground state, thereby emitting a photon. The photon is of just the correct energy to excite a transition from the ground state to the excited state in the absorber. This excited state then decays back to the ground state and re-emits a photon of the same energy.

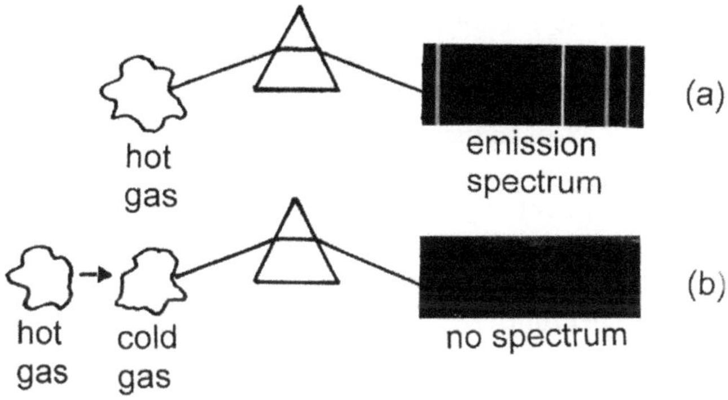

Figure 1.3. An atomic resonance fluorescence experiment. (a) Emission spectrum produced by the source of hot gas is observed by the spectrometer, and (b) emission spectrum from the source is absorbed by the absorber and greatly reduces the spectrum observed by the spectrometer.

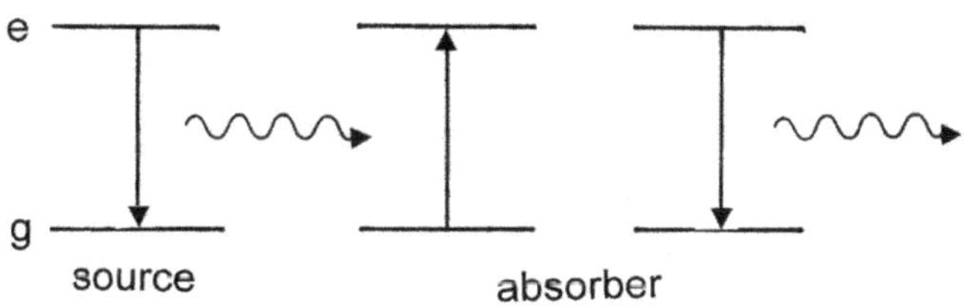

Figure 1.4. Atomic energy levels and transitions in the source and absorber during atomic resonance fluorescence; g = ground state and e = excited state.

The re-emitted photon is given off in a random direction (i.e., isotropically), thus decreasing the flux of photons in the direction towards the spectrometer. This phenomenon is fairly straightforward to observe, and is a clear demonstration of the quantized nature of the atomic energy levels.

Nuclei, like atoms, have well defined energy levels. This is because a nucleus is a bound system of particles (i.e., the neutrons and protons, collectively referred to as nucleons) just as the atom is a bound system of the nucleus and its electrons. Because both the neutron and the proton are fermions, they must occupy the energy levels in the nucleus in a way that satisfies the Pauli exclusion principle, just as the electrons in the atom must obey this principle. Thus, the lowest energy configuration of neutrons and protons in their energy levels that satisfies the Pauli principle represents the ground state of the nucleus, analogous to the situation for electrons in the atom. If a neutron and/or proton acquires energy and moves to a higher energy level, then the nucleus can be raised to an excited state. It is important to note that the energy levels of the nucleus are defined by the overall energy of the system and these energies are determined by the overall configuration of the nucleons in their energy levels. The energy levels of the nucleus are not really the same as the energy levels that the neutrons and protons can occupy but are the net result of the total configuration of particles. The same is true for atomic energy levels and electronic energy levels, although here the distinction is not as important as it is for nuclei.

The growing awareness of nuclear structure during the middle of the 20th century suggested the possibility of extending atomic resonance fluorescence experiments to consider an analogous phenomenon for nuclei, i.e., nuclear resonance fluorescence. Before we consider this possibility, it is important to look at the question of when two photons have the 'same' energy in more detail. Answering this question requires a detailed consideration of the Heisenberg uncertainty principle and the conservation of energy and momentum.

1.3 The Heisenberg linewidth and recoil energy

The monochromatic radiation that is emitted during a transition between quantum states is only monochromatic to a point. At a minimum, one must consider the distribution of energies associated with the unstable state in the context of the Heisenberg uncertainty principle. This principle states that the energy width associated with a state which has a mean lifetime of τ, is

$$\Delta E = \frac{\hbar}{2\tau} \tag{1.1}$$

(Frequently lifetimes of unstable states are given in terms of the half-life, $\tau_{1/2}$, where $\tau_{1/2} = (\ln 2)\,\tau$.) Equation (1.1) applies to the unstable excited state; the stable ground state has an infinite lifetime and, therefore, has zero energy width. The energy of the photon that is emitted by the source in figure 1.4 only needs to be the same as the energy that is absorbed by the absorber within a factor given by equation (1.1). If the atomic energy levels in the source and the absorber are the same, then this broadening of the energy will help to ensure that the absorber is able to absorb the photon that is emitted by the source.

Another factor that affects the energy of the photons is recoil. As energy and momentum must be conserved during the emission and absorption processes, a careful analysis of recoil is necessary. This analysis was first provided by Heinrich Kuhn (1904–94) in 1929. If the emitting atom is initially at rest, then, when the photon is emitted in one direction, the atom must recoil in the opposite direction in order to conserve linear momentum. The momentum of the photon is given by E_γ/c so conservation of momentum requires that the recoil momentum of the atom will be

$$p_R = E_\gamma/c.$$

The (non-relativistic) recoil energy of the atom can be written as

$$E_R = \frac{p_R^{\,2}}{2m} = \frac{E_\gamma^{\,2}}{2mc^2} \tag{1.2}$$

where m is the mass of the atom.

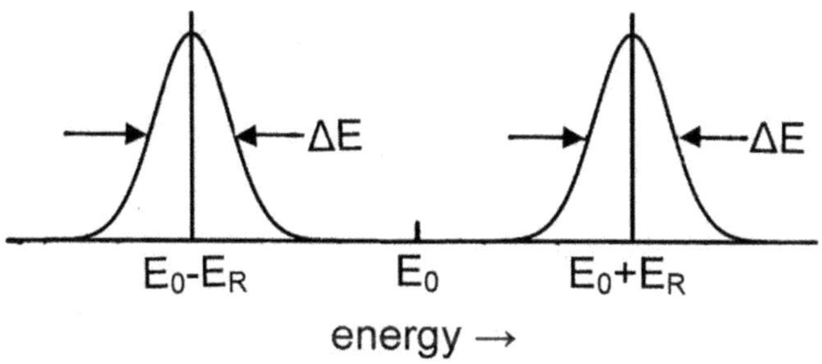

Figure 1.5. Energies of emitted and absorbed photons showing the importance of recoil energy and the energy linewidth given by the Heisenberg uncertainty principle.

Since conservation of energy requires that the energy of the photon plus the recoil energy of the atom must equal the energy available in the transition from the excited state to the ground state, E_0, then

$$E_0 = E_\gamma + E_R.$$

Combining the above equations gives the actual energy of the photon as

$$E_\gamma = E_0\left[1 - \frac{E_0}{2mc^2}\right]$$

where we have made the (valid) approximation that $E_\gamma \approx E_0$.

The presence of recoil means that the energy of the emitted photon is less than the energy available in the transition by the recoil energy of the atom. This also means that the photon energy that is required in order for the photon to be absorbed by the absorber atom will be the transition energy plus the atomic recoil energy. This situation is illustrated in figure 1.5. In the illustration the relative energies satisfy the relation $E_R > \Delta E$ and there is negligible overlap between the emitted and absorbed energies. Hence, in this particular example, atomic resonance fluorescence will not occur. It is important to consider some actual physical values in equations (1.1) and (1.2) to see if this is, indeed, the case. This calculation involves knowing (1) the lifetime of the excited state, (2) the energy of the excited state and (3) the mass of the atom. While all of these factors can vary considerably, we can look at some typical values. Using $\tau = 10^{-6}$ s, $E_0 = 2$ eV and $m = 50$ u, we find

$$\Delta E = 3.3 \times 10^{-10}\,\text{eV}$$

and

$$E_R = 4.3 \times 10^{-11}\,\text{eV}.$$

This calculation shows that figure 1.5 is not a good representation of the actual situation in a typical atomic resonance fluorescence experiment as the Heisenberg linewidth is an order of magnitude greater than the recoil energy. This means that the emission and absorption peaks align almost perfectly, giving rise to the successful observation of this resonance fluorescence phenomenon. The next section describes attempts to extend this approach to transitions between nuclear energy levels.

1.4 The early history of nuclear resonance fluorescence

Excited nuclear energy levels, which are the result of excitations of neutrons and/or protons in the nucleus, can decay back to the ground state and release a photon in the process. In the case of nuclear processes, the photons which are emitted are referred to as gamma rays, even though their energy may be similar to photons in the x-ray regime which are the result of atomic processes. We might expect that nuclear resonance fluorescence, analogous to atomic resonance fluorescence, might be observed. If we repeat the calculation from equations (1.1) and (1.2) for the nuclear case, we obtain some interesting results. Again, as in the case of atomic transitions, there is a wide range of values for the parameters needed for these calculations. However, we can use the following typical values: $\tau = 10^{-6}$ s, $E_0 = 10^5$ eV and $m = 50$ u. These values give the following results,

$$\Delta E = 3.3 \times 10^{-10}\ \text{eV} \tag{1.3}$$

and

$$E_R = 0.11\ \text{eV}.$$

Thus, we see that the Heisenberg width is the same as for the atomic case, but the recoil energy is many orders of magnitude larger. The condition that $E_R > \Delta E$ applies and the emission and absorption peaks do not overlap. The reason for this behavior is the very large value of the gamma ray energy associated with nuclear transitions, compared with the much smaller energy associated with atomic transitions.

In order to observe nuclear resonance fluorescence, we must either shift the location of the peaks closer to one another in order to get the peaks to overlap, or broaden the peaks substantially beyond the Heisenberg linewidth. Around the middle of the 20th century both approaches were being considered. In the first case, the source and absorber were moved relative to one another so that the relativistic Doppler shift moved the peaks closer together. In the second case, the source or absorber was heated in order to broaden the energy distribution of the peaks.

In order to Doppler shift the energy of one of the peaks (say the absorption peak) so that it coincides with the emission peak, it must be shifted by twice the recoil energy. Since the Doppler shift energy for a source moving with a relative velocity v is

$$E_D = \frac{E_0 v}{c}$$

then we require (using equation (1.2))

$$\frac{E_0 v}{c} = \frac{E_0^2}{mc^2}$$

or a velocity

$$v = \frac{E_0}{mc}. \tag{1.4}$$

Using $E_0 = 10^5$ eV and $m = 50$ u, gives a velocity of 640 m s^{-1}. One method of maintaining a velocity of this magnitude in the laboratory is to use an absorber in the form of a tilted rotating disk. If the disk is tilted by 45°, then a 1 m diameter disk must be rotated at about 17 000 rpm in order to compensate for the recoil energy. This is possible, but not easy.

The other approach is to heat the source or the absorber or both. Thermal energy will cause the atoms to move at random in the material, so that atoms in the source and absorber will sometimes be moving towards one another and will sometimes be moving apart. Atoms which absorb while they are moving toward the emitting atom in the source will have their absorption peak shifted towards the emission peak and absorbing atoms which are moving away from the emitting atom in the source will have their absorption peak shifted away from the emission peak. In this way, the emission and absorption peaks will be Doppler broadened as a result of thermal energy.

In the case where both the source and absorber are heated, the peaks will overlap significantly when the thermal broadening is comparable to the recoil energy, or

$$2k_B T = \frac{E_0^2}{2mc^2} \tag{1.5}$$

or for a temperature of

$$T = \frac{E_0^2}{4k_B mc^2}. \tag{1.6}$$

The factor of 2 on the left-hand side of equation (1.5) comes from directional averaging. Using the typical values from the above derivation gives $T = 620$ K (or about 350 °C). This is not a difficult temperature to maintain in the laboratory and thermal broadening was the most appropriate method of observing nuclear resonance fluorescence. In fact, Doppler shift experiments which are conducted at room temperature benefit from the room temperature thermal broadening of the transitions. Both approaches as described above provide similar insight into the ability to observe these effects. As can be seen from equations (1.4) and (1.6), our choice of the specific nuclear transition can significantly improve our ability to observe nuclear resonance fluorescence. Both equations show that it is advantageous to choose nuclei that are as massive as possible and to choose specific transitions with as little energy as possible. Both of these criteria are (more or less) consistent with one another as the typical spacing of nuclear energy levels decreases as nuclei become more massive. Looking back to equation (1.1) we also see that the use of transitions with short lifetimes may also be advantageous, although this will be discussed in more detail later in the book.

1.5 Problems

1.1. The transition from the first excited state to the ground state of ^{57}Fe (energy 14.4 keV, lifetime 141 ns) is used for a Mössbauer effect experiment. Calculate the Heisenberg linewidth of the transition.

1.2. A free ^{57}Fe nucleus emits a 14.4 keV gamma ray.
  (a) Calculate the recoil energy associated with this event.
  (b) Calculate the Doppler velocity necessary to completely compensate for the recoil.
  (c) Estimate the temperature needed to provide a substantial resonance fluorescence effect.

Bibliography and suggestions for further reading

Greenwood N N and Gibb T C 1971 *Mössbauer Spectroscopy* (London: Chapman and Hall)
Kalvius M and Kienle P (ed) 2012 *The Rudolf Mössbauer Story: His Scientific Work and Its Impact on Science and History* (Heidelberg: Springer)
Wertheim G 1964 *Mössbauer Effect: Principles and Applications* (New York: Academic)
Yoshida Y and Langouche G (ed) 2013 *Mössbauer Spectroscopy: Tutorial Book* (Heidelberg: Springer)

IOP Publishing

The Mössbauer Effect (Second Edition)

Richard A Dunlap

Chapter 2

The Mössbauer effect

2.1 Introduction

The experimental approaches as described above represent the situation at the time that Mössbauer began his doctoral studies. Mössbauer's experiments involved looking at the temperature dependence of the nuclear resonance fluorescence for the 129 keV first excited state to ground state transition in ^{191}Ir. It turns out that the large mass of the ^{191}Ir nucleus is advantageous for a significant nuclear resonance fluorescence effect, however, the transition energy is greater than that which is now generally considered acceptable. As well, the half-life of the 129 keV state is about 3.9 ns, which is less than that which is now commonly used for such experiments. However, at the time Mössbauer did his experiments, this half-life was not known, and it was expected that obtaining this value would be a benefit of the research. ^{191}Ir is rarely used for Mössbauer effect experiments now as there are other elements which are more interesting (as discussed below) and which have transitions that are easier to study. Mössbauer's original work was conducted in the decade following World War II and radioactive isotopes (even those which had no obvious military use) were heavily regulated and many more interesting alternatives to ^{191}Ir were not available for research use at that time.

2.2 Discovery of the Mössbauer effect

Because of the short half-life of the 129 keV state and the large mass of the ^{191}Ir nucleus, the overlap of the emission and absorption peaks was fairly favorable, despite the large transition energy. Figure 2.1 shows (to scale) the actual situation at room temperature for the 129 keV transition in ^{191}Ir. It was expected that, as the temperature was increased, the resonant effect would increase due to the increasing Doppler broadening of the peaks. It is generally difficult to observe a resonant effect, unless one is able to create a situation in which there is no resonant effect for comparison. In the case of ^{191}Ir, a source containing ^{191}Ir in the first excited state was used in conjunction with a gamma ray detector. A sample of ^{191}Ir was placed

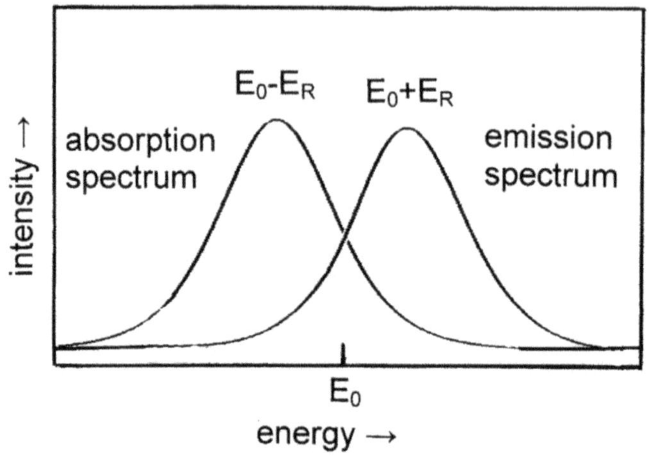

Figure 2.1. Actual room temperature relationship between the emission and absorption peaks of the 129 keV transition in ^{191}Ir.

(as the absorber) between the source and the detector and the decrease in the gamma flux was observed. For a non-resonant comparison, a piece of Pt was used between the source and detector. Pt has virtually the same non-resonant absorption as Ir (due to the similar atomic mass) but shows no resonant effect because the nuclear energy levels are different.

Figure 2.2 shows the temperature dependence of the absorption of 129 keV gamma rays from a ^{191}Ir source for an Ir absorber compared with a Pt absorber as measured by Mössbauer. As anticipated, the absorption increases slightly as the temperature is raised from around 280 K to 350 K as a result of the increased overlap of the emission and absorption peaks caused by thermal broadening. However, as Mössbauer lowered the temperature below room temperature, he saw a very substantial increase in the absorption, as shown in the figure. This behavior is the result of what we now refer to as the Mössbauer effect.

In order to better observe the details of the resonance effect, Mössbauer used an arrangement similar to that shown in figure 2.3 to scan through the resonance by Doppler shifting the gamma ray energy from the ^{191}Ir source. The energy of the gamma rays from the source could be shifted by a controlled amount by rotating the disk, either clockwise (for negative velocities) or counterclockwise (for positive velocities) and the gamma ray flux could be averaged over a period of time for different velocity settings.

The results of Mössbauer's experiment are illustrated in figure 2.4 and clearly show the presence of nuclear resonance fluorescence. The vertical axis is the percent difference in the counts per unit time from the condition where there is no resonant effect. The horizontal axis is shown in units of the Doppler velocity of the source. The energy shift corresponding to a particular velocity is given by the relativistic Doppler shift as

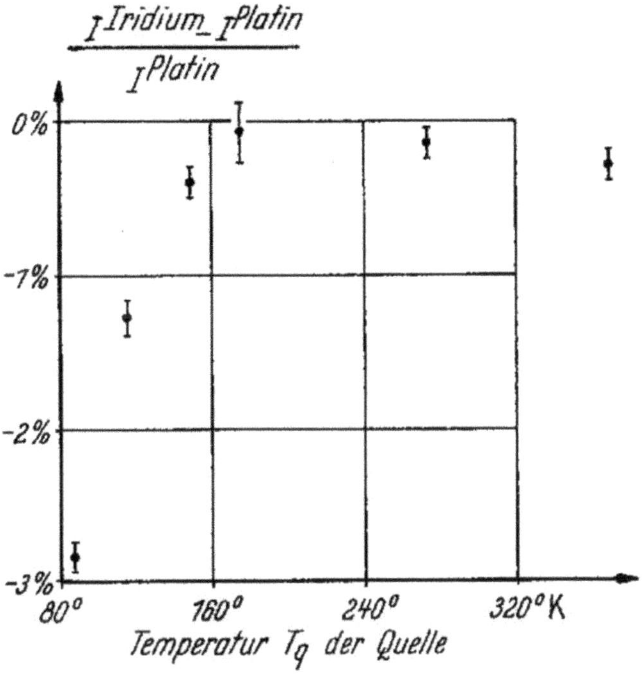

Figure 2.2. Temperature dependence of the absorption cross-section of ^{191}Ir at 129 keV as originally reported by Mössbauer (1958b). The relative change for ^{191}Ir compared to platinum is plotted and increasing absorption indicated by an increasing percentage in the negative direction on the vertical axis. Reprinted from Mössbauer (1958b), copyright (1958) with permission of Springer.

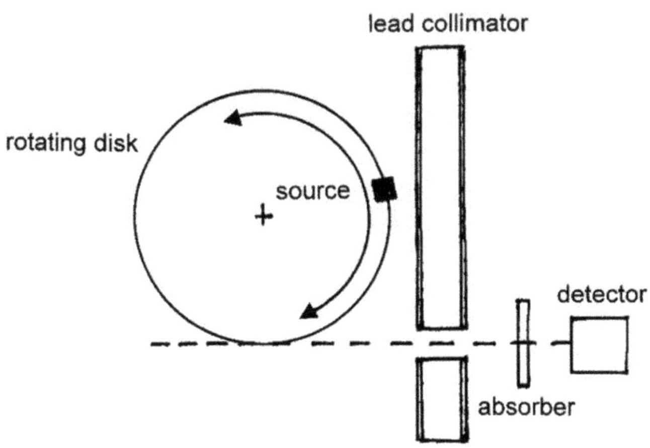

Figure 2.3. Schematic diagram of a simple Mössbauer effect experiment.

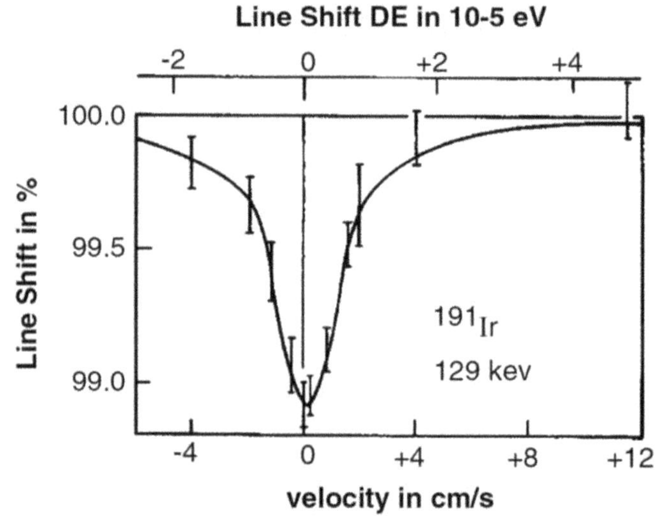

Figure 2.4. Resonant absorption as a function of Doppler velocity (or Doppler energy (DE) shift) for ^{191}Ir. Reprinted from Mössbauer (2012b), copyright (2012) with permission of Springer.

$$E - E_0 = \frac{v}{c} E_\gamma. \tag{2.1}$$

There are two obvious unusual features exhibited by the results in figure 2.4 that need to be addressed:

(1) Why is the maximum absorption at zero Doppler velocity if there is an energy shift due to recoil?
(2) Why is the resonance so narrow in energy (velocity) if thermal broadening is present?

The answers to both of these questions is that the resonance that Mössbauer observed is the result of transitions in which there is no recoil, that is, recoil-free nuclear resonance fluorescence. The basic reason for this behavior stems from the fact that the ^{191}Ir nuclei are in Ir atoms that are part of a crystalline lattice. In an atomic lattice, atoms cannot move freely but are constrained by the interactions they experience with their neighboring atoms.

One way of looking at this problem is based on the classical description of a crystal. An atom is constrained to be in its proper lattice site in the crystal by the interactions holding it there. When the nucleus of an atom emits a gamma ray by undergoing a transition to a lower energy level, then the conservation of energy and momentum requires that the atom will recoil. But as the atom is fixed in its relationship to its neighboring atoms and all the other atoms in the crystal, then it is the entire crystal that recoils and not the individual atom. If we now go back to equation (1.2) for the recoil energy, then the mass in the denominator on the right-hand side is not the mass of the atom but the mass of the entire crystal. Even if the

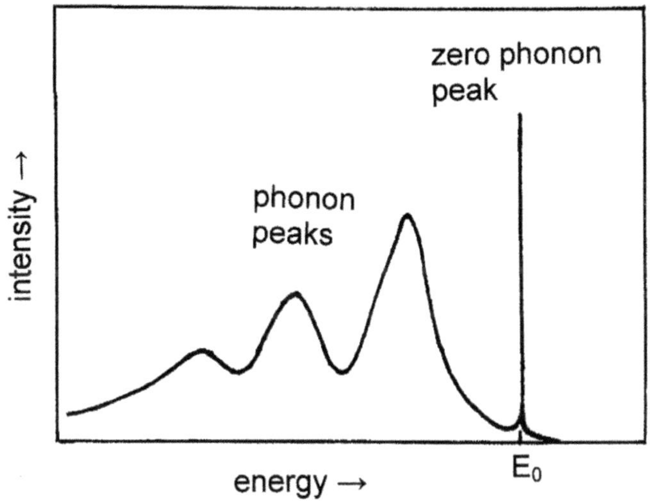

Figure 2.5. Gamma ray spectrum of source with zero, one, two... phonons being created.

sample is very small (say 1 μg) it would contain about 10^{20} atoms. Thus, the recoil energy calculated from equation (1.2) would not be 0.11 eV, as in equation (1.3), but about 10^{-21} eV, more than 11 orders of magnitude less than the Heisenberg linewidth in equation (1.3). So, for all practical purposes, there is no recoil and the emission and absorption peaks fall exactly on top of one another.

A more correct picture of this phenomenon requires a quantum mechanical view. If the nucleus of an atom emits a gamma ray as a result of a nuclear decay, then it can only move by emitting a phonon, that is a lattice vibration with quantized energy. We can view this process in a statistical way. If the recoil energy is less than the energy of the lowest energy phonon allowed in the material, then it is not likely that a phonon will be emitted, however, it is not impossible. Statistically, the phonon may or may not be created as long as the statistical average of such events conserves energy. If the recoil energy is greater, relative to the phonon energy, then phonon creation becomes more likely. If the recoil energy is greater still, there is the possibility of creating two or even more phonons. A typical gamma ray energy spectrum as produced by a source in a crystalline lattice is illustrated in figure 2.5. There is a sharp peak (defined by the Heisenberg linewidth, and some crystallographic effects as described later) where no phonon is created and there is no decrease of the gamma ray energy as a result of recoil. These emissions are referred to as zero-phonon events. There are gamma rays with lesser energy, as shown by the broad peaks to the left of the zero-phonon peak in the figure, which are emitted in events that produce one or more phonons, referred to as one-phonon events, two-phonon events, etc as appropriate. It is the total energy of the gamma ray and the phonon(s) that is conserved.

Figure 2.5 can be redrawn for nuclear emission and absorption from nuclei in atoms that are part of a crystal, as in figure 2.6. This diagram explains Mössbauer's results as shown in figure 2.4. It is only the zero-phonon events which overlap

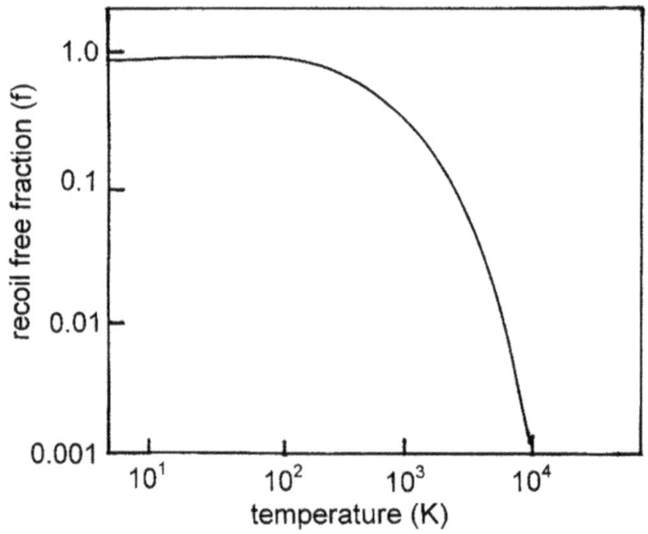

Figure 2.6. Temperature dependence of the recoil-free fraction for the 14.4 keV Mössbauer transition in ^{57}Fe.

perfectly at E_0 that are responsible for the resonance seen in the spectrum where no Doppler shift is required to align the emission and absorption spectra. It is also the zero-phonon events for which no thermal motion of the atoms contributes to the broadening of the spectral lines. Recoil-free nuclear resonance fluorescence (that is resonance fluorescence where the emissions and absorptions are zero-phonon events) is now known as the Mössbauer effect.

2.3 More about the Mössbauer effect

The above quantum mechanical description explains the observed spectrum in figure 2.4. However, we need to look a little deeper at the properties of phonons to fully explain the behavior shown in figure 2.2. As was indicated above, all gamma emissions and absorptions are not recoil-free. The fraction of transitions that are recoil-free (i.e. the recoil-free fraction, f) depends on both the energy associated with the nuclear transition and the behavior of phonons in the particular crystalline material which contains the nuclei. The behavior of phonons in a particular material depends on the nature of the chemical bonds in the material and also on the temperature. In general, we want to express the recoil-free fraction (in the source for example) as the probability that a nucleus in the excited state will decay to the ground state, thus emitting a photon, and in the process the lattice will not change state. This problem can be separated spatially into two regimens, the behavior on a scale of 10^{-15} m which describes what happens to the nucleus and the behavior on a scale of 10^{-10} m which describes what happens to the lattice. In order to calculate the probability that a photon will be emitted but the lattice will not change state we need to only deal with the large-scale behavior. In general, we can write the transition probability in Dirac notation as

$$f = \langle L_f | e^{ik\cdot r} | L_i \rangle \tag{2.2}$$

where L_i and L_f are the initial and final states of the lattice, respectively, and $e^{ik\cdot r}$ is a plane wave representing the emitted photon. In the case where the emission is recoil-free, then the initial and final states of the lattice are the same, so $L_f = L_i$. In order to solve this problem in detail, it is necessary to be able to mathematically express the properties of the lattice, including the energy and temperature dependence of the phonon spectrum. There are several models that can be used for this purpose, but the details are really not so important for the present discussion, as all reasonable models produce basically similar results. In general, the solution for the recoil-free transition probability as shown in equation (2.2) with $L_f = L_i$ has the form

$$f = K \exp(-2W) \tag{2.3}$$

where K is a constant and W is given by

$$W = \frac{E_\gamma^2}{mc^2} F'\left(\frac{T}{\theta_D}\right). \tag{2.4}$$

Here T is temperature, θ_D, is the Debye temperature and F' is a function that depends on the details of the lattice model. The Debye temperature is a measure of the stiffness of the bonds between the atoms in a material and is related to the energies associated with phonons in the material. In general, the Debye temperature scales, more or less, with the melting temperature of the solid.

Equations (2.3) and (2.4) allow us to see the general features of the temperature dependence of the recoil-free fraction. As W decreases with decreasing temperature, then equation (2.3) shows that the recoil-free fraction (as shown in figure 2.6) will increase as the temperature is lowered. It is this behavior of the recoil-free fraction that is responsible for the changes in the resonant absorption that are depicted in figure 2.2.

A final feature of figure 2.4 for discussion at this point is the shape of the resonance curve. Using an experimental set-up as shown in figure 2.3 to scan Doppler velocities (and hence energies) is equivalent to measuring the overlap of the recoil-free source and absorber energy curves as shown in figure 2.7 as they are moved past one another. In an ideal case, this yields a resonance curve that is the same shape as the energy distribution of the excited state that results from the Heisenberg uncertainty principle. The decay of an unstable state is a process that occurs exponentially in time, that is, if there are N_0 nuclei in the excited state at time $t = 0$, then the number of nuclei remaining in the excited state as a function of time will be

$$N(t) = N_0 \exp(-t/\tau) \tag{2.5}$$

where τ is the mean lifetime of the excited state. When we measure the resonance spectrum, as in figure 2.4, we measure the energy dependence of the resonant cross-section. The relationship between the response of the system in the energy domain

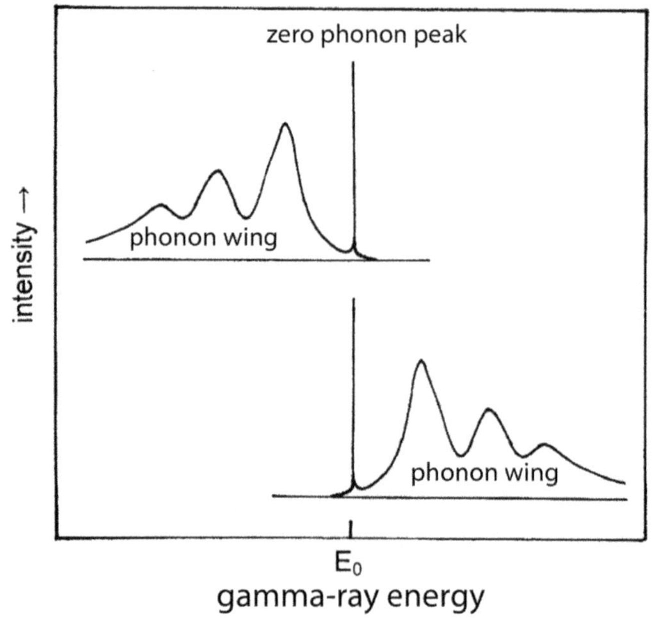

Figure 2.7. Gamma-ray emission spectrum (top) and absorption spectrum (bottom) showing zero-phonon events with no recoil energy and events which create phonons.

and the response of the system in the time domain is given by a Fourier transform. The Fourier transform of the exponential decay function in equation (2.5) is a Lorentzian, $L(v)$. The theoretical functional form of the Mössbauer spectrum is given in velocity units, as defined in equation (2.1), by the Fourier transform of equation (2.5) as

$$L(v) = \frac{A}{\Gamma^2 + (v - v_0)^2} \tag{2.6}$$

where A/Γ^2 is the maximum amplitude of the resonance above background, v_0 is the center of the resonance curve and Γ is the half width at half maximum of the resonance curve in velocity units. This is the basic functional form that is almost always used to analyze Mössbauer effect spectra and is discussed in more detail in section 8.6.

2.4 Choice of a Mössbauer transition

Since Mössbauer's original observation of recoil-free nuclear resonance fluorescence in ^{191}Ir, dozens of other (mostly better) isotopes of numerous elements have been shown to exhibit the same phenomenon. We have seen a couple of criteria previously with respect to a classical view of this effect that could be used to gauge the suitability of a transition for the observation of the Mössbauer effect. However, it is now apparent that from a quantum mechanical perspective these criteria are mainly aimed at maximizing the recoil-free fraction. In a number of cases, isotopes do not have a sufficient recoil-free fraction at room temperature, but experiments can be successfully conducted by lowering the temperature of the source and/or absorber.

The relevant criteria for a successful Mössbauer experiment may be summarized as follows:

A large nuclear mass. This is important to ensure a large recoil-free fraction as indicated in equations (2.3) and (2.4).

A small transition energy. This is even more important than the large mass as indicated by the functional form of equation (2.4).

A reasonable excited state half-life. This will determine the width of the resonance, as seen in figure 2.4. If the half-life is too short the resonance will be very broad. This will reduce the ability to analyze complex spectra as discussed in later chapters. If the half-life is too long, then the resonance will be very sharp and controlling the velocity during the measurement can be problematic.

A reasonable Debye temperature. This is a function of the material in which the Mössbauer nucleus is contained and not a direct function of the nuclear properties. A low Debye temperature, particularly in conjunction with other factors that are not ideal, will make the measurement more difficult and may require cooling the source and/or absorber.

High natural abundance of the relevant isotope. This allows experiments to be performed on materials that are not isotopically enriched. This has the advantage that experiments are easier and less expensive, and it allows experiments to be performed on naturally occurring materials such as biological materials and geological materials.

Appropriate nuclear spin states. The details of the nuclear spin will determine how a nucleus will interact with electric and magnetic fields in its surroundings. This will be discussed in detail in chapters 4–7.

An acceptable precursor. It is necessary to have a source of characteristic gamma rays from the source. As the half-life of the excited state is typically a fraction of a microsecond, this state would normally be populated by the decay of a radioactive precursor. Thus, the ability to perform traditional Mössbauer effect experiments depends on the availability of a suitable precursor. It is also important that the source may be prepared in a suitable form. Most of the remainder of this book deals with the means by which the Mössbauer effect can be utilized for the study of the solid-state properties of materials. As such, it is important that the characteristics of the source do not significantly affect the quality of the experimental results. This aspect of Mössbauer effect spectroscopy is discussed further in the next three sections and in chapter 8. It should be noted, as well, that Mössbauer experiments can be performed in-beam at accelerators that produce short-lived nuclides or by using synchrotron radiation and these techniques are discussed in chapter 11.

The most commonly used Mössbauer nuclide is ^{57}Fe. The second most commonly used nuclide is ^{119}Sn. In the next two sections the use of ^{57}Fe and ^{119}Sn for

Mössbauer effect experiments is discussed in some detail. The final section of this chapter briefly reviews some of the other Mössbauer nuclides that have been frequently utilized.

2.5 Properties of ^{57}Fe

Although no nuclide perfectly satisfies all the criteria given above, the 14.4 keV transition in ^{57}Fe represents one of the best compromises. As well, iron is a common element with many practical applications. It commonly occurs in biological systems, in mineral specimens and has commercial applications because of its structural, magnetic and electrical properties. Table 2.1 summarizes some of the most important properties of the 14.4 keV transition in ^{57}Fe with respect to its use in Mössbauer spectroscopy. The significance of the values of the parameters in table 2.1 is discussed below.

Mass. Although the mass, ~57 u, is not as great as many other potential Mössbauer isotopes, it is a less important factor in equation (2.4) than some other criteria, such as transition energy.

Energy. The transition from the first excited state to the ground state of ^{57}Fe has an energy of 14.4 keV. This is substantially smaller than most nuclear transition energies and, as equation (2.4) indicates, this is the single most important factor in determining the recoil-free fraction.

Half-life. The half-life of the 14.4 keV state of ^{57}Fe is about 99 ns. This is sufficiently long that the resonance line is well defined and allows for considerable information to be extracted from a Mössbauer measurement, but is short enough that velocity control is not unnecessarily complex.

Debye temperature. Although this is a feature of the lattice and not the nuclear transition itself, iron generally bonds with other atoms to produce alloys and compounds that have a reasonably high Debye temperature and, hence, a fairly high melting point. This means that from a recoil-free standpoint most iron Mössbauer experiments can be conducted at room temperature. Some iron containing organometallic materials are more difficult in this respect.

Table 2.1. Properties of ^{57}Fe relevant to its use as a Mössbauer nuclide.

Property	Value
Mass	56.935 u
Energy	14.4 keV
Half-life	99.3 ns
Natural abundance	2.1%
Nuclear spin states	$3/2^- \rightarrow 1/2^-$
Precursor	^{57}Co
Precursor half-life	270 d

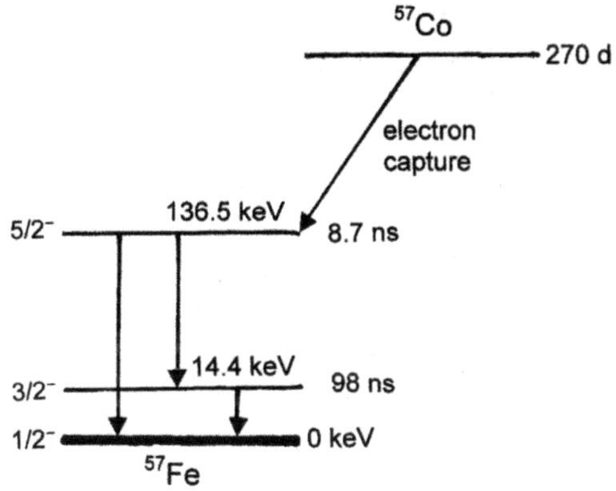

Figure 2.8. Energy level diagram showing the electron capture decay of ^{57}Co to ^{57}Fe and the gamma decay of excited ^{57}Fe nuclear states. Nuclear spins for the ^{57}Fe energy levels are shown in the figure.

Natural abundance. The natural abundance of ^{57}Fe is 2.1%. The remaining iron atoms in natural iron are ^{54}Fe (5.9%), ^{56}Fe (91.72%) and ^{58}Fe (0.28%). This is the one factor where iron's properties fall short of ideal. However, in cases where factors such as Debye temperature or iron concentration in the sample make the experiment difficult, there may be the (expensive) option of using iron enriched in ^{57}Fe.

Nuclear spin states. The 14.4 keV state of ^{57}Fe has a spin of 3/2 and the ground state has a spin of 1/2. This is an ideal situation as it optimizes the sensitivity of the transition to the physical properties of the material without unnecessarily complicating the details of the spectrum. This will be discussed further in chapters 4–7.

Precursor. As the half-life of the excited state of ^{57}Fe is about 98 ns, we needed a means of continuously populating this level throughout the duration of the experiment. For atomic resonance fluorescence experiments photon energies are a few eV so thermal energy or an electric discharge can be used to populate an excited atomic state from the ground state. Even a relatively small nuclear excited state energy, as for ^{57}Fe, is too large to use these methods. Instead, it is customary to populate excited states by the nuclear decay of a precursor nucleus. The method of doing this for ^{57}Fe is shown in figure 2.8. The precursor nucleus, ^{57}Co, has 27 protons and 30 neutrons. It decays with a half-life of about 270 days by electron capture;

$$e^- + {}^{57}\mathrm{Co} \rightarrow {}^{57}\mathrm{Fe} + \mathrm{n} + \nu_e.$$

This decay process converts a proton to a neutron by capturing an atomic electron,

$$e^- + \mathrm{p} \rightarrow \mathrm{n} + \nu_e$$

where ν_e is an electron neutrino. The by-product of this decay is the nucleus of a ^{57}Fe atom with 26 protons and 31 neutrons. About 99% of the time the electron capture process leaves the ^{57}Fe nucleus in an excited state at 136 keV. This state decays either to the ^{57}Fe ground state (11% of the time) or to the 14.4 keV state (89% of the time). By this method, the 14.4 keV state of ^{57}Fe is continuously populated and a ^{57}Co source will serve as a source of 14.4 keV ^{57}Fe Mössbauer gamma rays for an extended period of time. As ^{57}Co is an unstable nuclide, it must be synthesized by an appropriate method. The most common technique for preparing ^{57}Co involves the irradiation of ^{56}Fe (which is about 92% naturally abundant) with deuterons by the reaction

$$d + {}^{56}Fe \rightarrow {}^{57}Co + n. \tag{2.7}$$

The cross-section for this reaction is favorable for deuterons with an energy of 9.5 MeV. The ^{57}Co that is used as a source of 14.4 keV γ-rays for Mössbauer spectroscopy must be in a crystallographic environment where it has a large recoil-free fraction and is unaffected by crystalline disorder and not subject to perturbations from hyperfine interactions, as discussed in chapters 5 and 7. Palladium is a common host material for a ^{57}Co Mössbauer source, but Cr, Cu and Rh, are also used.

2.6 Properties of ^{119}Sn

Table 2.2 summarizes some of the most important properties of the 23.8 keV transition in ^{119}Sn with respect to its use in Mössbauer spectroscopy. The significance of the values of the parameters in table 2.2 is discussed below.

Mass. This is an improvement over ^{57}Fe and contributes to an acceptable recoil-free fraction.

Energy. The transition energy for ^{119}Sn is somewhat greater than for ^{57}Fe and approximately negates the advantages of the greater mass.

Half-life. The half-life of the 23.8 keV state of ^{119}Sn is about 18 ns. This is not as ideal as the somewhat longer half-life of the 14.4 keV state of ^{57}Fe. Although this results in a somewhat greater natural Heisenberg linewidth, it is still suitable for detailed Mössbauer studies of solid-state properties.

Debye temperature. Elemental tin has a relatively low Debye temperature as evidenced by its low melting point (232 °C). Tin in alloys and compounds

Table 2.2. Properties of ^{119}Sn relevant to its use as a Mössbauer nuclide.

Property	Value
Mass	118.903 u
Energy	23.8 keV
Half-life	18.3 ns
Natural abundance	8.6%
Nuclear spin states	$3/2^+ \rightarrow 1/2^+$
Precursor	^{119m}Sn
Precursor half-life	250 d

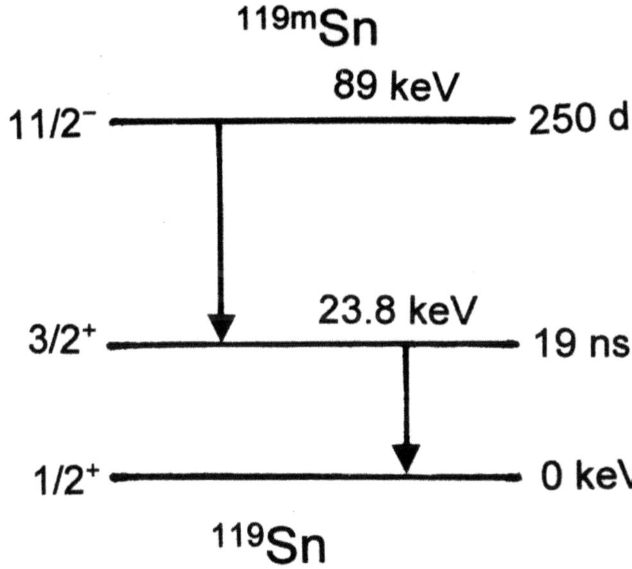

Figure 2.9. Energy level diagram showing the gamma decay of ^{119m}Sn. Nuclear spins for the energy levels are shown in the figure.

often has higher Debye temperatures. Typically, ^{119}Sn Mössbauer measurements can be made at room temperature even with elemental tin (e.g., see section 8.5).

Natural abundance. Tin has the largest number of naturally occurring isotopes (10) of any element because of its particularly stable proton configuration. ^{120}Sn at about 33% is the most naturally abundant isotope. ^{119}Sn, at 8.6% natural abundance, is an improvement in this respect over ^{57}Fe.

Nuclear spin states. The 14.4 keV state of ^{57}Fe has a spin of 3/2 and the ground state has a spin of 1/2. This is an ideal situation as it optimizes the sensitivity of the transition to the physical properties of the material without unnecessarily complicating the details of the spectrum. This will be discussed further in chapters 4 to 7.

Precursor. The 23.8 keV Mössbauer state of ^{119}Sn is populated by a γ-decay from a metastable state, ^{119m}Sn, as illustrated in figure 2.9. ^{119m}Sm can be prepared by neutron irradiation of 24% naturally abundant ^{118}Sn by the reaction

$$n + {}^{118}Sn \rightarrow {}^{119m}Sn + \gamma.$$

$CaSnO_3$, $BaSnO_3$ and vanadium provide a suitable chemical environment and good Debye temperature and are common hosts for ^{119m}Sn sources. As with ^{57}Co sources, the half-life of ^{119m}Sn allows for the use of the source for an extended period of time.

2.7 Other common Mössbauer nuclides

The Mössbauer effect has been observed in more than 40 different elements as shown in the periodic table in figure 2.10. In many elements, the Mössbauer effect has been observed in more than one isotope of the element and in some cases more than one transition in a particular isotope. Suitable Mössbauer transitions are quite common in the heavier elements because of the typically more closely spaced nuclear energy levels.

Figure 2.10 provides an indication of how commonly different elements have been studied by means of Mössbauer effect spectroscopy. Specific nuclides and transitions are in common use because the Mössbauer parameters facilitate the measurement and/or the element or nuclide is of particular scientific or commercial interest. Table 2.3 gives information on some nuclides that are in common use for Mössbauer spectroscopy. More extensive lists of Mössbauer nuclides and their properties are provided by Greenwood and Gibb (1971) and Shenoy and Wagner (1978).

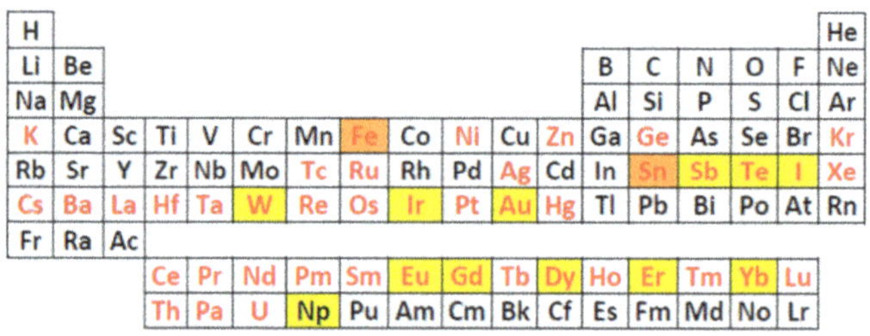

Figure 2.10. Mössbauer effect periodic table showing Mössbauer active elements in red. The two most commonly used elements for Mössbauer, Fe and Sn, are highlighted in orange. Other elements that are in fairly common use for Mössbauer experiments are highlighted in yellow. From Principi (2020) CC BY 4.0

Table 2.3. Properties of some common Mössbauer transitions.

	Nuclide				
Property	^{121}Sb	^{125}Te	^{155}Gd	^{193}Ir	^{197}Au
Mass	120.904 u	124.905 u	154.923 u	192.963 u	196.967 u
Energy	37.17 keV	35.48 keV	86.54 keV	73.0 keV	77.34 keV
Half-life	3.46 ns	1.48 ns	6.5 ns	6.09 ns	1.91 ns
Natural abundance	57.4%	7.12%	14.8%	62.7%	100%
Nuclear spin states	$7/2^+ \rightarrow 5/2^+$	$3/2^+ \rightarrow 1/2^+$	$5/2^+ \rightarrow 3/2^-$	$1/2^+ \rightarrow 3/2^+$	$1/2^+ \rightarrow 3/2^+$
Precursor	^{121m}Sn	^{125}I	^{155}Eu	^{193}Os	^{197}Pt
Precursor decay mode	β^-	ec	β^-	β^-	β^-
Precursor half-life	55 y	59.4 d	4.76 y	30.5 h	18.3 h
Precursor preparation	^{120}Sn(n,γ)	^{123}Sb(α,2n)	^{154}Sm(n,γ)	^{192}Os(n,γ)	^{196}Pt(n,γ)

An investigation of the properties given in table 2.3 and a comparison with the properties of ^{57}Fe (in particular) and ^{119}Sn provides some insight into the various approaches that need to be taken in order to conduct successful Mössbauer spectroscopy experiments.

Mass. In all cases given in the table, the mass is greater than that of ^{57}Fe. In fact, this is the case for virtually all Mössbauer nuclides. ^{40}K is, perhaps, the only exception.

Energy. In all cases the transition energy is substantially greater than that for ^{57}Fe. As a result, the use of these Mössbauer nuclides would benefit greatly by cooling the source and/or the absorber to low temperatures to improve the recoil-free fraction. In fact, this is required for nuclides with transitions greater than about 70 keV in the table.

Half-life. The short half-lives of the Mössbauer state for the nuclides in the table means that the Heisenberg linewidth will be larger than for ^{57}Fe. In some cases, this may limit the ability to resolve certain spectral features.

Natural abundance. In all cases, the natural abundances are somewhat or significantly greater than for ^{57}Fe.

Spin states. In three cases, the spin states are equivalent to those of both ^{57}Fe and ^{119}Sn, meaning that the spectral features will be similar to those of these two nuclides. However, ^{121}Sb and ^{155}Gd have transitions involving higher spin states. As discussed in detail in chapters 5 and 7, the resulting Mössbauer spectra will be more complex and this, together with the greater linewidth, may make data analysis more difficult.

Precursor. Both ^{121}Sb and ^{155}Gd Mössbauer utilize sources with long half-lives and, as for ^{57}Fe and ^{119}Sn, a commercially obtained source will be suitable for an extended period of time. The half-life of the ^{125}Te precursor puts some limitations on the longevity of the source. However, ^{193}Ir and ^{197}Au precursors have half-lives that are comparable to a typical Mössbauer measurement. These, as well as many Mössbauer nuclides with similarly short-lived precursors, are only feasible at research locations that have access to facilities where the source can be prepared by the reaction in the table. In the case of ^{193}Ir and ^{197}Au, source preparation is by the (n,γ) reaction using thermal neutrons. Thermal neutron sources are fairly common at larger universities and their availability makes such Mössbauer experiments feasible. The use of other nuclides with suitable Mössbauer transitions but very short-lived precursors is discussed in chapter 11.

While the Mössbauer effect began as a bit of a nuclear curiosity, it soon became apparent that the potential applications of this phenomenon were enormous. The usefulness of Mössbauer effect spectroscopy comes primarily from the fact that nuclei will interact with external (to the nucleus) electric and magnetic fields and that the Mössbauer effect is an extremely sensitive method of observing these interactions. In order to understand the ways in which Mössbauer spectroscopy has been put to use, we need to begin by understanding a little more about the properties of the nucleus.

2.8 Problems

2.1. (a) What is the relevance of each of the parameters for the magnitude of the recoil-free nuclear resonance fluorescence?

(i) the energy of the gamma ray;

(ii) the mass of the atom;

(iii) the energy of the phonons in the lattice.

(b) For each parameter above, is it advantageous to maximize or minimize the parameter value?

2.2. ^{61}Ni is sometimes used for Mössbauer effect spectroscopy. Locate relevant information about this nuclide and discuss the potential difficulties associated with this experiment.

2.3. For an ideal Mössbauer transition, derive an expression relating the parameter Γ in equation (2.6) and the lifetime of the Mössbauer state.

Bibliography and suggestions for further reading

Greenwood N N and Gibb T C 1971 *Mössbauer Spectroscopy* (London: Chapman and Hall)

Kalvius M and Kienle P (ed) 2012 *The Rudolf Mössbauer Story: His Scientific Work and Its Impact on Science and History* (Heidelberg: Springer)

Mössbauer R L 1958a Kernresonanzabsorption von Gammastrahlung in Ir-191 *Naturwissenschaften* **45** 538–9

Mössbauer R L 1958b Kernresonanzabsorption von Gammastrahlung in Ir-191 *Z. Phys.* **151** 124–43

Mössbauer R L 2012a Recoilless nuclear resonance absorption of gamma radiation *The Rudolf Mössbauer Story: His Scientific Work and Its Impact on Science and History* ed M Kalvius and P Kienle (Heidelberg: Springer) pp 3–17

Mössbauer R L 2012b The discovery of the Mössbauer effect *The Rudolf Mössbauer Story: His Scientific Work and Its Impact on Science and History* ed M Kalvius and P Kienle (Heidelberg: Springer) pp 37–48

Principi G 2020 The Mössbauer effect: a romantic scientific page *Metals* **10** 992

Shenoy G K and Wagner F E (ed) 1978 *Mössbauer Isomer Shifts* (Amsterdam: North-Holland)

Yoshida Y and Langouche G (ed) 2013 *Mössbauer Spectroscopy: Tutorial Book* (Heidelberg: Springer)

Richard A Dunlap

Chapter 3

Properties of the nucleus

3.1 Introduction

The interaction between a nucleus and external electric and magnetic fields is described in terms of the quantum mechanical spin of the nucleus and the electric and magnetic fields that the nucleus produces. The present chapter considers the way in which the quantum numbers of the nucleus are determined and the ways in which the electric and magnetic fields produced by the nucleus are described.

3.2 Nuclear quantum numbers

Because electrons are fermions, the Pauli exclusion principle governs their arrangement in the atomic energy levels. Since neutrons and protons are also fermions, their configuration in the nuclear energy levels must also comply with these same criteria. There are, however, two fundamental differences between these two situations that must be considered. Firstly, neutrons and protons are different particles, and each must satisfy the Pauli principle independently and, secondly, the potential which binds protons and neutrons in the nucleus is a combination of the strong interaction and (for protons) the Coulomb interaction, while electrons in atoms are subject to only the Coulomb interaction.

The neutron and proton energy levels are quantized and are designated by an ordinal quantum number, n, and the orbital angular momentum quantum number, l. The quantum number n is not the principal quantum number, as it is in the case of atomic energy levels, but is merely an index that counts the number of energy levels with a particular value of l. As well, there is no requirement for nucleons in a nucleus, as there is for electrons in an atom, that $l < n$. Contrary to the case for atomic energy levels where m_l (i.e., the z-component of l) is a good quantum number, this is not the case for the nucleons in a nucleus. This feature is the result of the presence of a strong spin–orbit coupling for nucleons. The presence of this coupling means that l and s are no longer independent quantum numbers but that

doi:10.1088/978-0-7503-6039-5ch3

they couple to yield a total spin, j, for each nucleon, where the value of j is constrained to

$$j = l \pm s.$$

For the neutron and the proton, the intrinsic spin is $s = 1/2$, yielding

$$j = l \pm 1/2.$$

Thus, j and its z-component, m_j, are independent (good) quantum numbers for nucleons in the nucleus, giving a set of four quantum numbers, n, l, j and m_j, that must satisfy the Pauli exclusion principle for each type of nucleon. The total spin of the nucleus, $\boldsymbol{I}$, is the vector sum of the spins of all the neutrons, $\boldsymbol{J}_\mathrm{n}$, and all protons, $\boldsymbol{J}_\mathrm{p}$;

$$\boldsymbol{I} = \boldsymbol{J}_\mathrm{n} + \boldsymbol{J}_\mathrm{p}$$

where the total spin of the neutrons and total spin of the protons are given as the vector sum of the spins of the individual nucleons,

$$\boldsymbol{J}_n = \sum_i^N \boldsymbol{j}_{ni} \tag{3.1}$$

and

$$\boldsymbol{J}_p = \sum_i^Z \boldsymbol{j}_{pi} \tag{3.2}$$

respectively. Because of the strong pairing tendency for nucleons in the nucleus, total neutron or proton spins in equations (3.1) and (3.2) are generally the result of a single unpaired nucleon.

3.3 Electromagnetic multipole moments of the nucleus

To understand the concept of nuclear electromagnetic moments, we begin with a simple picture of how the nucleus produces electric and magnetic fields. The nucleus contains neutrons and protons. It is easy to see that the charged protons will produce an electric field and, if they are moving, a magnetic field. It is not so easy to see why uncharged neutrons may be important. However, the neutron is made up of charged quarks and this feature may be responsible for the production of a magnetic field. We begin here with a discussion of electric fields where only the protons are relevant.

The simplest configuration of electric charges is the electric monopole or an isolated (point) charge. This is illustrated in figure 3.1. The magnitude of the electric field is given by

$$E = \frac{q}{4\pi\varepsilon_0 r^2}$$

where q is the charge, ε_0 is the permittivity of free space and r is the distance from the charge. The charge q is referred to as the electric monopole moment. It is, therefore,

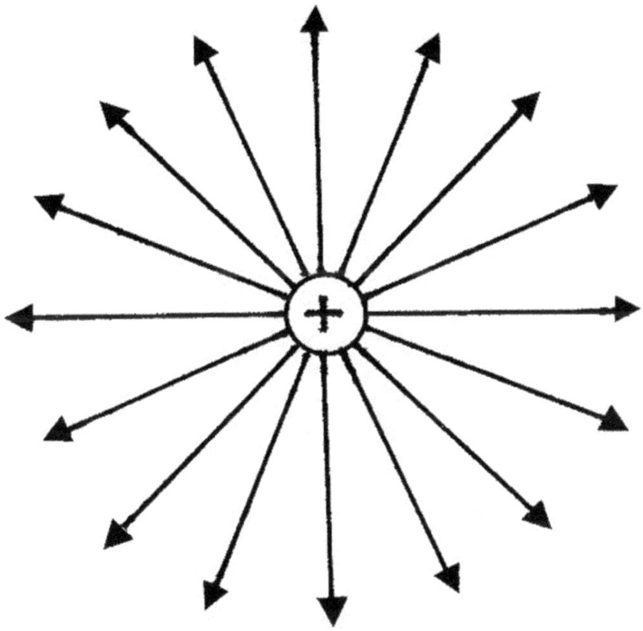

Figure 3.1. The electric monopole (point charge) showing the electric field lines.

clear that a nucleus will have a net electric monopole moment by virtue of the positive charges associated with the protons.

The electric dipole as shown in figure 3.2 is comprised of a positive charge and a negative charge (in the simple case, the two charges are of the same magnitude) separated by a distance $\boldsymbol{a}$. The details of the electric field lines will be related to the magnitude of the charges and the distance between them. It is convenient to define an electric dipole moment, $\boldsymbol{p}$, where

$$\boldsymbol{p} = q\mathbf{a}.$$

This is defined as a vector quantity as the distance $\boldsymbol{a}$ may be defined as a vector pointing along the direction between the charges from positive to negative.

The next configuration of charges that is of interest is the electric quadrupole, as shown in figure 3.3. This configuration consists of alternating positive and negative charges (all the same magnitude) arranged on the corners of a square. The electric quadrupole moment may be described as a vector

$$\boldsymbol{Q} = qA$$

where $\boldsymbol{A}$ is a vector with a magnitude equal to the area of the square in a direction normal to the plane containing the charges.

The nucleus is not a point charge or a simple geometric arrangement of charges, as described above, but is best described as a spatial distribution of charges defined in terms of a charge distribution, $\rho(\boldsymbol{r})$. Note, that at this point, we are making no assumptions about the symmetry (spherical or not) of this distribution. The

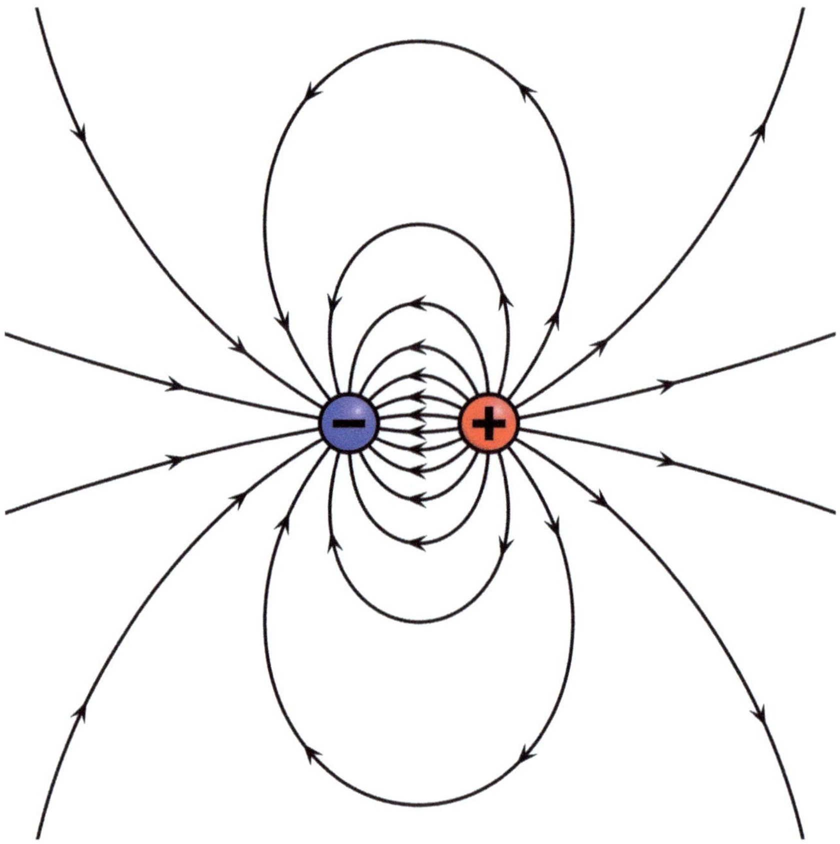

Figure 3.2. The electric dipole showing the electric field lines. This [Electric dipole field of two point-charges with finite distance] image has been obtained by the author from the Wikimedia website where it was made available by Geek3 (2010a) under a CC BY-SA 3.0 licence. It is included within this chapter on that basis. It is attributed to Geek3.

interaction between the nucleus and external electric fields can be viewed in terms of the energy associated with the nuclear charge distribution in a slowly (spatially) varying electric potential, $\phi(\boldsymbol{r})$. It is appropriate to consider a slowly varying potential as the external electric field will arise (as we will see later) from the charges associated with electrons in the material. This assumption is valid since the spatial extent of the electron charge distribution is commensurate with the size of the atom (i.e., 10^{-10} m) while the spatial extent of the charge distribution of the nucleus is typically about 10^{-15} m.

We write the interaction energy as described above as

$$U = \int \rho(\boldsymbol{r})\phi(\boldsymbol{r})d^3\boldsymbol{r}. \tag{3.3}$$

Since $\phi(\boldsymbol{r})$ is a slowly varying function of $\boldsymbol{r}$, it is appropriate to express this function as a Taylor expansion about the origin;

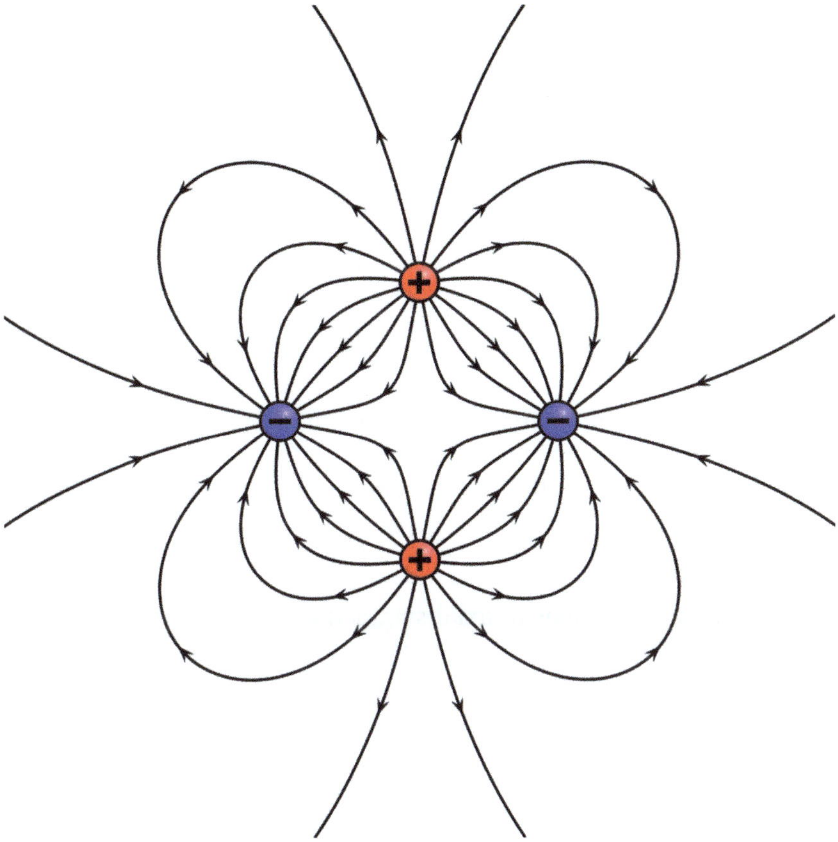

Figure 3.3. The electric quadrupole showing electric field lines. This [Field of two positive and two negative point-charges constituting a non-ideal electric quadrupole] image has been obtained by the author from the Wikimedia website where it was made available by Geek3 (2010b) under a CC BY-SA 3.0 licence. It is included within this chapter on that basis. It is attributed to Geek3. Geek3 (2010b). Field of two positive and two negative point-charges constituting a non-ideal electric quadrupole https://commons.wikimedia.org/wiki/File:VFPt_four_charges.svg.

$$\phi(\vec{r}) = \phi(0) + (\boldsymbol{r} \cdot \nabla\phi)_{r=0} + \frac{1}{2}\left[\sum_{i,j} x_i x_j \frac{\partial^2 \phi}{\partial x_i \partial x_j}\right]_{r=0} + \cdots \tag{3.4}$$

where x_i and x_j are the Cartesian spatial coordinates, x, y, z. Substituting equation (3.4) into equation (3.3) gives the energy as

$$U = \phi(0)\int \rho(\boldsymbol{r})d^3\boldsymbol{r} + \left[\int \rho(\boldsymbol{r})\boldsymbol{r}d^3\boldsymbol{r}\right] \cdot \nabla\phi(\boldsymbol{r}) + \frac{1}{2}\sum_{i,j}\frac{\partial^2\phi}{\partial x_i \partial x_j}\int \rho(\boldsymbol{r})x_i x_j d^3\boldsymbol{r} + \cdots. \tag{3.5}$$

The first term on the right in equation (3.5) is the monopole interaction term, where the monopole moment is merely the integral of the charge distribution

$$q = \int \rho(\mathbf{r}) d^3\mathbf{r}. \tag{3.6}$$

The second term on the right in equation (3.5) is the dipole term, where the dipole moment is expressed as

$$P = \int \rho(\mathbf{r}) \mathbf{r} d^3\mathbf{r}. \tag{3.7}$$

The third term on the right in equation (3.5) is the quadrupole term, where the quadrupole moment is defined as

$$Q = \int \rho(\mathbf{r}) x_i x_j d^3\mathbf{r}. \tag{3.8}$$

Although we have made no assumptions about the symmetry of the nucleus, it is generally appropriate at this point to assume that the nucleus has axial symmetry (even if it is not spherical). In this case, equation (3.8) may be written as

$$Q = \int \rho(\mathbf{r}) [3z^2 - r^2] d^3\mathbf{r} \tag{3.9}$$

where the z-direction is the axis of axial symmetry.

We can now see some general features of the multipole moments of the nucleus. The monopole moment of the nucleus, as given by equation (3.6), is merely eZ where e is the magnitude of the electronic charge and Z is the atomic number (i.e., the number of positively charged protons in the nucleus). Unlike figure 3.2, the nucleus does not contain any free negative charges and a careful inspection of the electric dipole moment as defined by equation (3.7) shows that this quantity is zero. Equation (3.9) shows that if the positive charge distribution of the nucleus (from its protons) is spherically symmetric, then the electric quadrupole moment will disappear. If the positive charge distribution of the nucleus is distorted from spherical symmetry, even though it retains axial symmetry, then the electric quadrupole moment will be non-zero. A close inspection of equation (3.9) shows that if the charge distribution of the nucleus is flattened along the axis of symmetry, then the electric quadrupole moment will have a negative sign. On the other hand, if the charge distribution is stretched along the axis of symmetry, then the electric quadrupole moment will have a positive sign. This is discussed further in chapter 5.

We now turn to the magnetic moments of the nucleus. There is no evidence that magnetic monopoles exist, so they can be ignored in the present discussion. A classical magnetic dipole consists of a current loop as shown in figure 3.4 where, in this example, the current consists of an orbiting positive charge (as appropriate for the positively charged protons in the nucleus). The current produces a magnetic dipole moment, $\boldsymbol{\mu}$, where

$$\boldsymbol{\mu} = \frac{1}{2} q \mathbf{r} \times \mathbf{v} = \frac{1}{2} q \omega r^2.$$

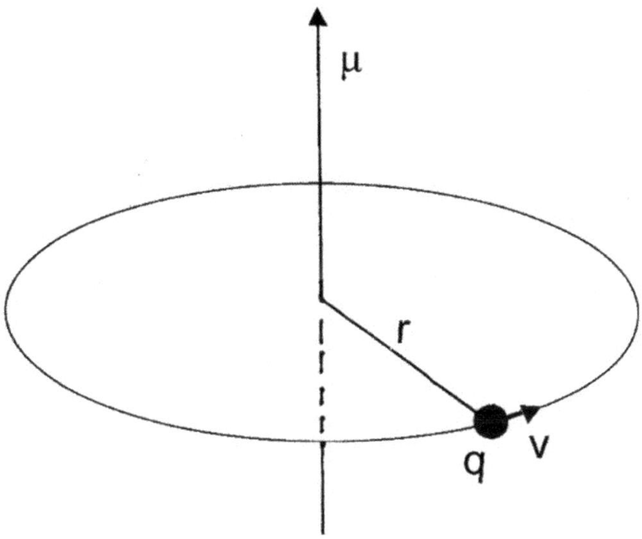

Figure 3.4. The magnetic dipole as a current loop.

Therefore, the magnetic dipole moment associated with the nucleus can arise as a result of charged protons in the nucleus. However, from a quantum mechanical standpoint, the formation of the magnetic dipole moment of the nucleus is somewhat more complex. Firstly, there is a dipole moment associated with the orbital component of the angular momentum of the proton for reasons as illustrated in figure 3.4. In addition, there is a component of the nuclear dipole moment that comes from the intrinsic spins of the protons in the nucleus. Finally, there is a component of the nuclear dipole moment that comes from the intrinsic spins of the neutrons. Although the neutron has no net charge, its internal structure consists of charged quarks and this gives rise to a magnetic moment associated with the neutron which can contribute to the dipole moment of the nucleus. Since the neutron has no net charge there is no corresponding orbital contribution to the nuclear moment from the neutron.

A magnetic quadrupole field may be produced by a combination of currents in solenoids, as shown in figure 3.5, where there is a minimum in the field at the center of the array. The nucleus can, in principle, possess a nuclear quadrupole moment, although in general this is small and is not relevant to the present discussion.

The above analysis of the electromagnetic multipole moments of the nucleus shows that there are three moments of the nucleus that need to be considered in an analysis of the interaction between the nucleus and any external electric or magnetic fields, the electric monopole moment (eZ), the electric quadrupole moment given by equation (3.9) and the magnetic dipole moment. These three electromagnetic multipole moments of the nucleus give rise to three well defined interactions with external electromagnetic fields that perturb the nuclear energy levels, and which can be observed by the features they produce in Mössbauer spectra as described in the following chapters.

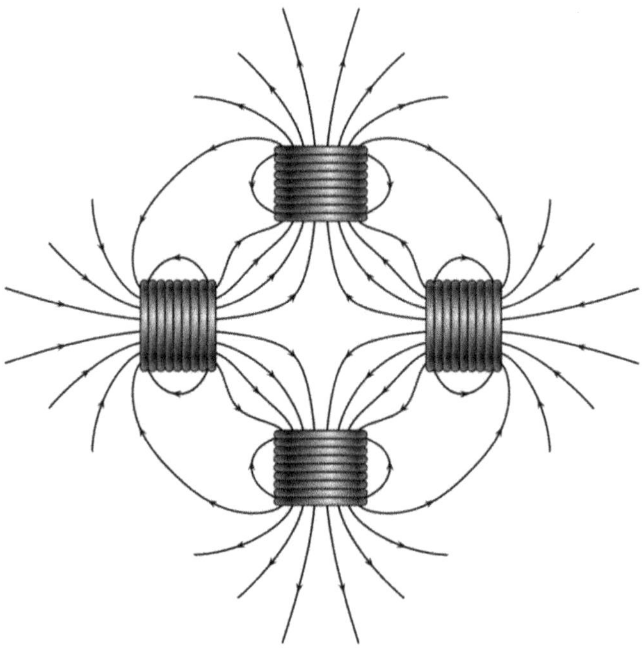

Figure 3.5. The magnetic quadrupole produced by currents in four solenoids. This [Magnetic field of four coils creating a quadrupole field] image has been obtained by the author from the Wikimedia website where it was made available by Geek3 (2010c) under a CC BY-SA 3.0 licence. It is included within this chapter on that basis. It is attributed to Geek3.

Bibliography and suggestions for further reading

Cottingham W N and Greenwood D A 1998 *An Introduction to Nuclear Physics* (London: Chapman and Hall)

Das A and Ferbel T 2003 *Introduction to Nuclear and Particle Physics* 2nd edn (Singapore: World Scientific)

Dunlap R A 2004 *An Introduction to the Physics of Nuclei and Particles* (Belmont, CA: Brooks/Cole)

Geek3 2010a *Electric Dipole Field of Two Point-Charges with Finite Distance* (https://commons.wikimedia.org/wiki/File:VFPt_dipole_electric.svg)

Geek3 2010b *Field of Two Positive and Two Negative Point Charges Constituting a Non-Ideal Electric Quadrupole* https://commons.wikimedia.org/wiki/File:VFPt_four_charges.svg

Geek3 2010c *Magnetic Field of Four Coils Creating a Quadrupole Field* https://commons.wikimedia.org/wiki/File:VFPt_quadrupole_coils_1.svg

Henley E M and Garcia A 2007 *Subatomic Physics* 3rd edn (Singapore: World Scientific)

Krane K S 1988 *Introductory Nuclear Physics* (New York: Wiley)

Williams W S C 1991 *Nuclear and Particle Physics* (Oxford: Clarendon)

Chapter 4

Hyperfine interactions—part I: the electric monopole interaction and the chemical isomer shift

4.1 Introduction

The interactions which are studied by means of the Mössbauer effect are the interactions between the charges in a material and the nuclear electromagnetic moments. These interactions are referred to as hyperfine interactions, because their effect on nuclear energy levels is extremely small. However, it is the exceptional energy resolution of the Mössbauer effect that makes this technique so powerful, and which allows for the measurement of these very small perturbations to the nuclear energy levels. We begin with the electric monopole interaction that gives rise to the perturbation of the nuclear energy levels known as the chemical isomer shift.

4.2 The electric monopole interaction

The interaction between the electronic charge distribution and the nuclear electric monopole moment results from the fact that the nucleus is not a point charge but has a spatial charge distribution that overlaps the electronic charge distribution from the atomic electrons. This interaction energy is expressed as

$$E = \int \rho(\boldsymbol{r})\left[\phi_{\text{real}}(\boldsymbol{r}) - \phi_{\text{point}}(\boldsymbol{r})\right]d^3r \tag{4.1}$$

where $\rho(\boldsymbol{r})$ is the electronic charge distribution, ϕ_{real} is the Coulomb potential due to the actual charge distribution of the nucleus and ϕ_{point} is the Coulomb potential from a point charge nucleus. The electronic charge distribution is rather complex, but we can understand some of the basic features of this function by looking at the theoretical charge distributions for different electronic states of a hydrogen-like

doi:10.1088/978-0-7503-6039-5ch4

atom. In general, the charge distribution for the electrons in an atom can be expressed in terms of the electron wavefunction, $\Psi(\mathbf{r})$, as

$$\rho(\mathbf{r}) = -e\,|\Psi(\mathbf{r})|^2.$$

The electron wavefunction is obtained from the solution to the Schrödinger equation in three dimensions,

$$\frac{\hbar^2}{2m}\nabla^2\Psi(\mathbf{r}) - V(\mathbf{r}) = -E\Psi(\mathbf{r})$$

where $V(\mathbf{r})$ is the Coulomb potential and E are the energy eigenvalues. It is assumed that, in spherical coordinates, the wavefunction is separable as

$$\Psi(r,\theta,\varphi) = R(r)\Theta(\theta)\Phi(\varphi)$$

and that the radial part of the wavefunction satisfies the radial equation

$$\frac{\hbar^2}{2m}\left[\frac{d^2R}{dr^2} + \frac{2}{r}\frac{dR}{dr}\right] - \left[V(r) + \frac{l(l+1)\hbar^2}{2mr^2}\right] = -ER. \tag{4.2}$$

Some solutions for equation (4.2) with a Coulomb potential are shown in figure 4.1 for different principal quantum numbers, n, and orbital angular momentum quantum numbers, l.

Since the nucleus, and hence the nuclear Coulomb potential is spatially very small compared to the electron charge distribution, it is only the portion of the radial wavefunction in figure 4.1 that is very close to the origin that contributes to the

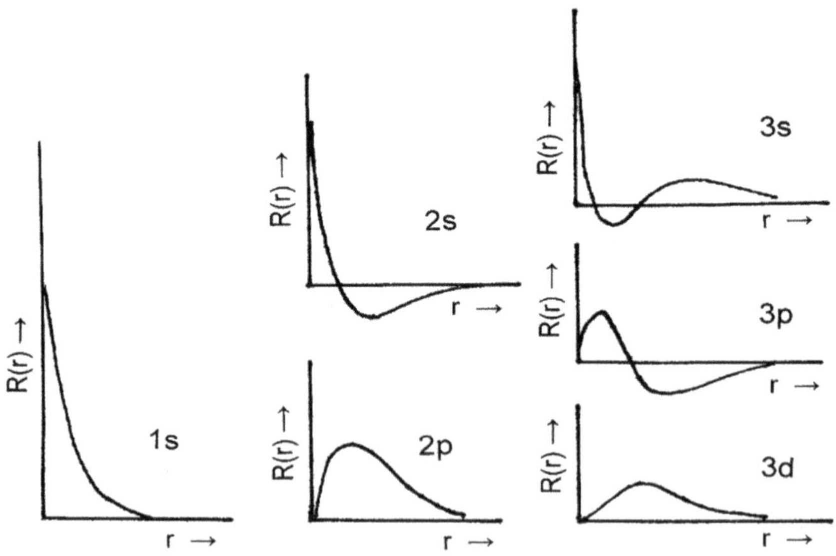

Figure 4.1. Radial part of the electron wavefunction for a hydrogen-like atom for different values of the principal (n) and orbital angular momentum (l) quantum numbers. All graphs are plotted on the same arbitrary scale.

charge distribution in equation (4.1). Figure 4.1 shows that it is only the s-state wavefunctions that have significant non-zero values near the origin, and which contribute to the electric monopole interaction energy.

The real nuclear Coulomb potential depends on the radial distribution of the nuclear charge distribution, that is, the distribution of protons in the nucleus, $\rho_p(r)$. This has been measured for many nuclei using high energy electron scattering experiments. A typical result from this type of experiment, where the charge distribution is assumed to be spherically symmetric, is illustrated in figure 4.2. The nuclear charge distribution consists of a fairly flat region near the center and a gradually decreasing region near the edge. While various mathematical models have been used to describe this shape, it is sufficient for the present derivation to use a simple step function, as shown in figure 4.3, where the nucleus is defined to have a radius of R_0.

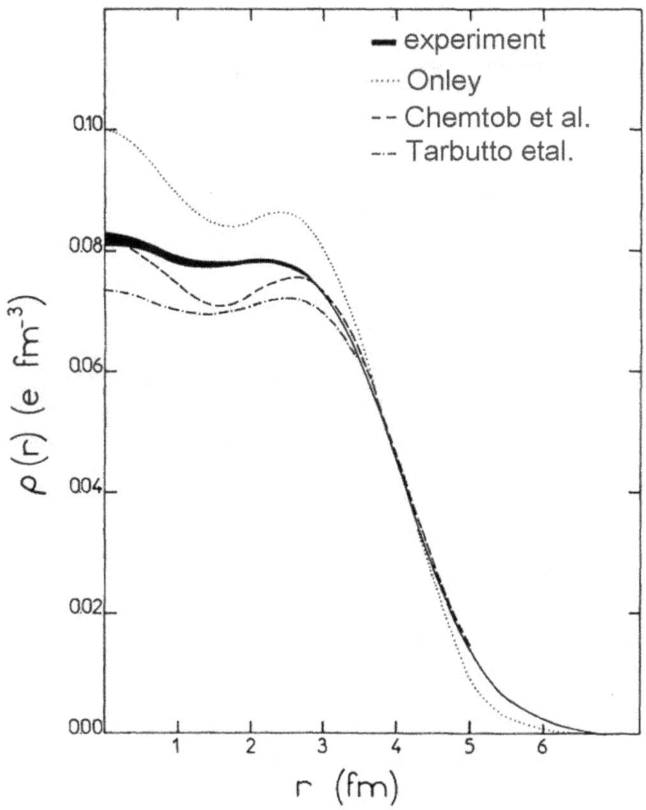

Figure 4.2. Measured nuclear charge distribution for a ^{58}Ni nucleus. The solid line is from the analysis of experimental data and the broken lines are from different theoretical models. Theoretical predictions are from Onley (1968), Chemtob *et al* (1974) and Tarbutton and Davies (1968), as noted in the figure. Reprinted with permission from Sick *et al* (1975), copyright (1975) by the American Physical Society.

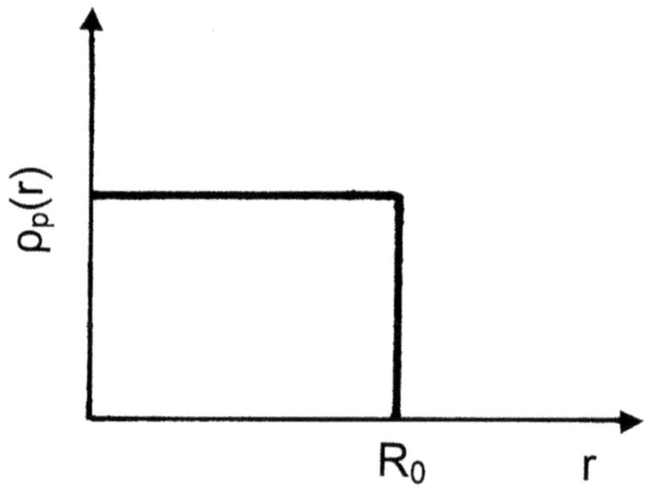

Figure 4.3. Simple step function used for the nuclear charge distribution.

Since the integral of the proton charge distribution, $\rho_p(r)$, over the size of the nucleus must equal the total charge of the nucleus, Ze, then the normalization condition for the charge distribution is

$$Ze = \int_0^{R_0} \rho_p(r) d^3 r.$$

This charge distribution gives a nuclear potential of the form

$$\phi_{\text{real}}(r) = \frac{Ze}{R_0}\left[\frac{3}{2} - \frac{r^2}{2R_0^2}\right]. \tag{4.3}$$

The nuclear potential associated with a nucleus that is a point charge is merely the Coulomb potential,

$$\phi_{\text{point}}(r) = \frac{Ze}{r}. \tag{4.4}$$

The above may be used to evaluate the interaction energy in equation (4.1) by realizing that the electronic wavefunctions vary on a distance scale that is about five orders of magnitude larger than the spatial scale for the variations of the nuclear potential. Thus, in equation (4.1) the electronic charge distribution may be considered a constant over the portion of the integral that contributes to the interaction energy. We designate the electronic charge density in terms of the wavefunction at the origin as

$$\rho(0) = -e\,|\Psi(0)|^2 \tag{4.5}$$

Substituting equations (4.3) to (4.5) into (4.1) gives

$$E = -\frac{2\pi Z e^2 R_0^2}{5}\,|\Psi(0)|^2. \tag{4.6}$$

The electric monopole interaction energy is a function of the square of the nuclear radius and the square of the electron wavefunction evaluated at the origin. The first of these quantities is a property of the nucleus (only) and not a property of the electronic environment. The second of these quantities is a function of the electronic environment which results from the properties of the atom (only) and is not a function of the properties of the nucleus. This distinction will become very important in the discussion below.

We now need to think about how this result relates to the way in which a Mössbauer effect experiment is conducted. Figure 4.4(a) shows a simple energy level diagram for a source and an absorber where our previous discussion showed that recoil-free resonance fluorescence would be observed, as in Mössbauer's experiment, with a center at zero Doppler shift energy. Figure 4.4(b) shows what happens when we include the effects of the electric monopole interaction. Each of the energy levels (i.e., ground and excited states) in the source and absorber, are shifted by the electric monopole interaction energy as given by equation (4.6). The ground and excited states differ because the properties of the nucleus (including its mean radius) are different in these two nuclear states. The states in the source and absorber differ because the electronic environment of the nuclei is not necessarily the same in these two materials. So according to equation (4.6), each of the four energy levels is shifted by a quantity that depends on the nuclear state (through R_0) and the electronic

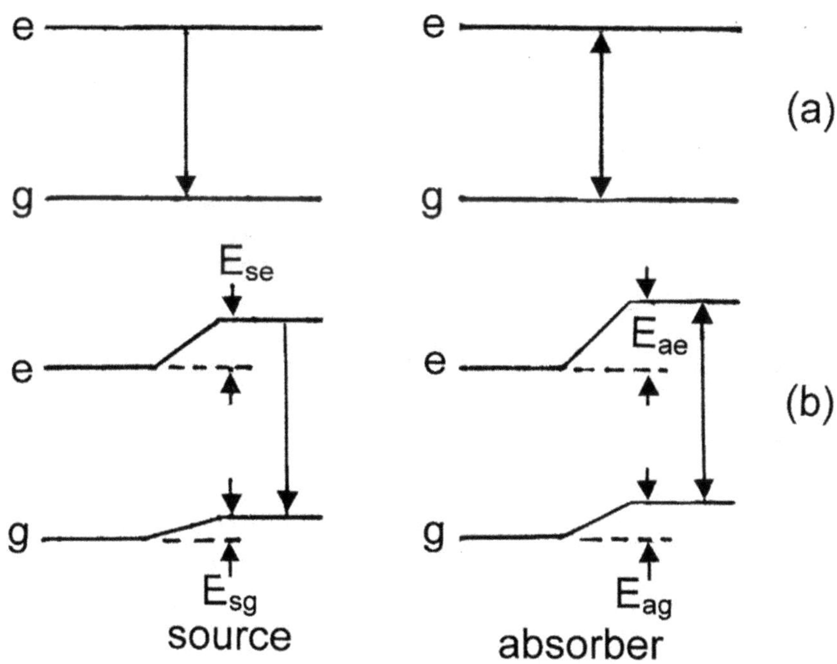

Figure 4.4. Energy levels in the source and absorber in a Mössbauer effect experiment for (a) the case where the electric monopole interaction is not considered and (b) the case where the electric monopole interaction is included showing energy shifts of the ground (g) and excited (e) states in the source (s) and absorber (a).

environment (through $\Psi(0)$). As these four energy shifts are, in general, different from one another, then there is a net difference in the transition energy in the source and the transition energy in the absorber,

$$\Delta E = (E_{ae} - E_{ag}) - (E_{se} - E_{sg}). \tag{4.7}$$

Combining equation (4.7) with equation (4.6), gives a net energy difference between the transition energy in the source and the transition energy in the absorber as

$$\Delta E = \frac{2\pi Z e^2}{5}\left[\,|\Psi_a(0)|^2 - |\Psi_s(0)|^2\,\right] \cdot \left[R_e^2 - R_g^2\right]. \tag{4.8}$$

Here, the first term in square brackets describes the difference in the electronic environment in the source and the absorber and the second term in square brackets gives the difference of the square of the nuclear radius in the ground and excited states. In a Mössbauer experiment, this energy shift is observed by the fact that a Doppler shift must be introduced to compensate for the electric monopole energy and the maximum absorption is no longer observed at zero velocity. A schematic of a Mössbauer effect spectrum illustrating this effect is shown in figure 4.5 where the velocity shift, referred to as the isomer shift (or sometimes the chemical isomer shift), δ, is related to the energy shift in equation (4.8) by the Doppler effect as,

$$\delta = \frac{c\Delta E}{E_\gamma}.$$

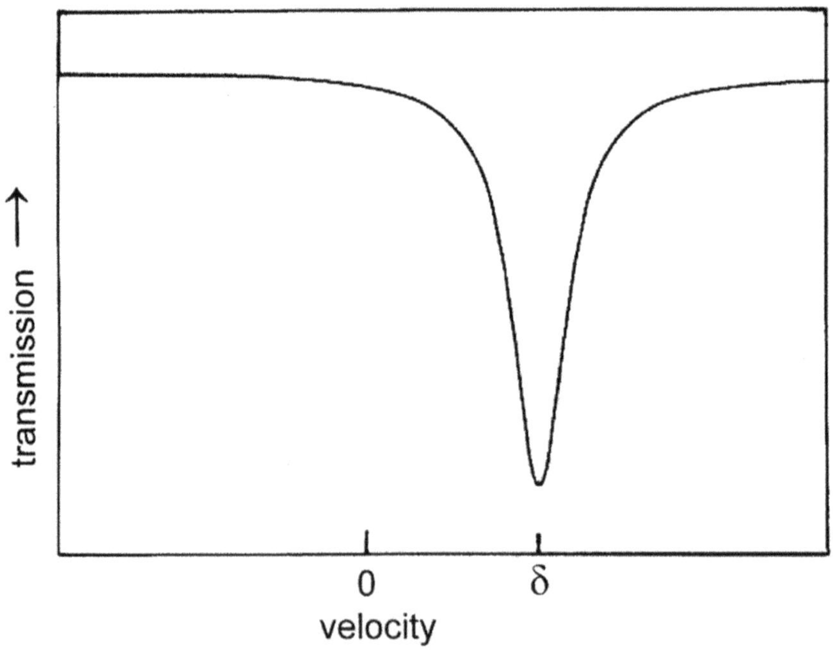

Figure 4.5. Isomer shift, δ, as observed in a Mössbauer effect experiment.

All Mössbauer effect experiments made with a particular nuclear transition will have the same value of the term $R_e^2 - R_g^2$ in equation (4.8) and this term may be considered a constant for the transition. If the term $|\Psi_s(0)|^2$ is known, or can be determined, then the experiment, as described above, is a convenient means of determining $|\Psi_a(0)|^2$, that is the electron density at the nuclei of the atoms in the absorber. We can look at a simple illustration to see how this can be useful. As the vast majority of Mössbauer research is conducted on the 14.4 keV transition in ^{57}Fe, we will consider this as an example in the next section.

4.3 The chemical isomer shift

One of the most useful aspects of Mössbauer effect spectroscopy, and the measurement of the isomer shift in particular, is the ability to understand the interactions between the ^{57}Fe probe atom and its environment. ^{57}Fe atoms in the absorber material will interact with their neighboring atoms through their chemical bonds, so we need to analyze how these chemical bonds will influence $|\Psi_a(0)|^2$, or, from equation (4.5), the electron density at the origin, $\rho(0)$. To do this, we need to look at the electronic configuration of different valence states of an iron atom.

A neutral iron atom has 26 electrons in the configuration $1s^2 2s^2 2p^6 3s^2 3p^6 3d^6 4s^2$. If some of the electrons contribute to the formation of chemical bonds, then the configuration of localized electrons will be altered. Table 4.1 shows the electronic configuration for free iron ions in different valence states. Note that, in general, the 4s electrons are less tightly bound than the 3d electrons and, therefore, at least for the free ion case, will be the first to be lost, although in solids this may not always be the case.

Figure 4.6 shows some typical experimental results relating the measured isomer shift to the iron valence state. Isomer shifts are measured relative to room temperature metallic α-Fe as discussed further in chapter 8. Note that there is a small temperature dependence to the isomer shift which is considered in the next section. There is some range of typical isomer shift values for each valence state, but the figure shows that there are very clear trends. We can understand these trends on the

Table 4.1. Electronic configuration for free iron ions in different valence states. Note: $[Ar] = 1s^2 2s^2 2p^6 3s^2 3p^6$.

Valence	Electron configuration
0	$[Ar]3d^6 4s^2$
1+	$[Ar]3d^6 4s^1$
2+	$[Ar]3d^6$
3+	$[Ar]3d^5$
4+	$[Ar]3d^4$
5+	$[Ar]3d^3$
6+	$[Ar]3d^2$

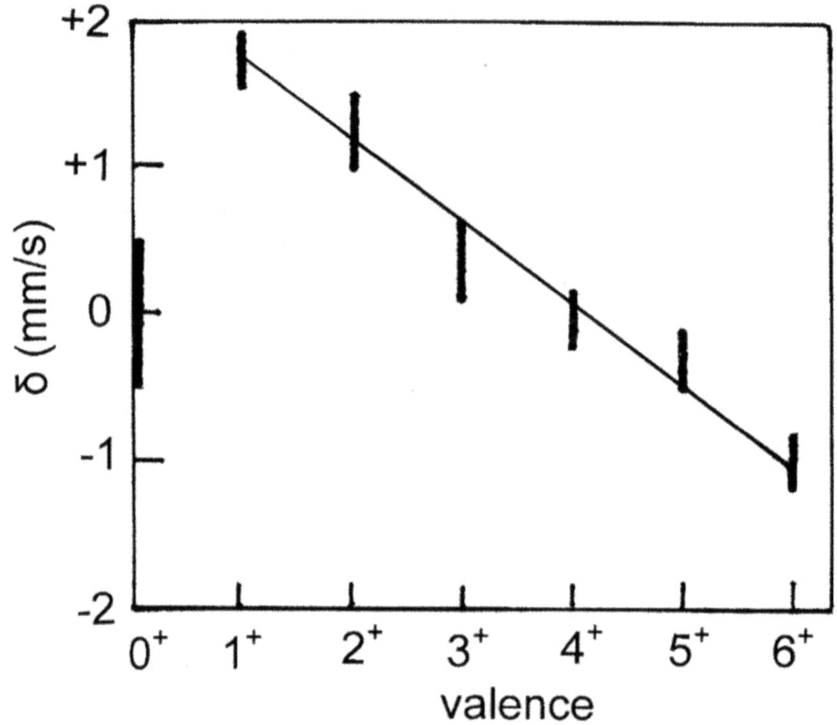

Figure 4.6. Relationship between ^{57}Fe Mössbauer isomer shift and known oxidation state for some iron containing materials. The vertical bars show the typical range of isomer shifts for each valence state.

basis of the electronic configurations shown in table 4.1. The total electron density at the nucleus will be the sum of the densities from the different s-state orbitals, that is

$$\rho(0) = \rho_{1s}(0) + \rho_{2s}(0) + \rho_{3s}(0) + \rho_{4s}(0). \tag{4.9}$$

It can be shown that changes in valence state will have a relatively small effect on the density distribution of the 1s and 2s electrons. If we consider the trends shown in figure 4.6 for valences 2+ to 6+, then these ions have no remaining 4s electrons, so an evaluation of equation (4.9) only needs to consider the effects of the 3s electrons. Specifically, we need to consider how changes in the number of 3d electrons will affect $\rho_{3s}(0)$ and hence the isomer shift. As the valence state increases from 2+ to 6+ the number of localized 3d electrons decreases from 6 to 2. A decreasing number of 3d electrons actually alters the 3s electron distribution so as to increase $\rho_{3s}(0)$. This is because the decrease in 3d electrons decreases the screening of the 3s electrons from the positive nuclear charge and allows them to be more strongly attracted to the nucleus. For ^{57}Fe, an increasing electron density at the nucleus results in a less positive isomer shift. This feature results from the fact that the term $[R_e^2 - R_g^2]$ in equation (4.8) is negative for ^{57}Fe, meaning that the radius of the 14.4 keV excited state of the nucleus is less than the radius of the ground state. The very large positive isomer shift observed in figure 4.6 for Fe^{1+} may be best explained if the electron configuration of this valence state for iron in an ionic solid is more like $3d^7$ rather than $3d^6 4s^1$, as one would expect for a free ion.

Perhaps the greatest significance of the results shown in figure 4.6, is the ability to use this information to determine the valence state of iron in compounds where the electronic configuration of the iron atoms is not known. This relationship will be considered in more detail in chapters 9 and 10 on the applications of the Mössbauer effect. It should be noted that, in all cases, the electric monopole interaction is present, so that Mössbauer spectra which show features of electric quadrupole or magnetic dipole interactions will also always show an isomer shift that is characteristic of the electric monopole interaction.

4.4 The second order Doppler shift

In a solid at a finite temperature, the atoms that are taking part in the Mössbauer effect are vibrating on their lattice sites. These vibrations have a characteristic frequency of the order of 10^{12} Hz. The time scale for a Mössbauer event is of the order of the lifetime of the Mössbauer state, which is 10^{-9} s. Thus, on the time scale of the Mössbauer event, the average displacement of the Mössbauer atoms will be zero. However, if we look more carefully at the Doppler effect, as described by equation (2.1) and consider relativistic effects, we find that this simple explanation of atomic vibrations is not sufficient.

The relativistic Doppler effect gives the observed frequency, f', of a photon emitted by a source moving at a velocity, v, as

$$f' = f\frac{\left(1 - \frac{v}{c}\right)}{\left(1 - \frac{v^2}{c^2}\right)^{1/2}}$$

where f is the frequency in the stationary system. This expression may be approximated as

$$f' = f\left(1 - \frac{v}{c}\right)\left(1 + \frac{v^2}{2c^2}\right). \tag{4.10}$$

The term v/c in first order term in equation (4.10) will, as noted above, average to zero over the Mössbauer event time. However, because v^2 is independent of direction, the second order term in equation (4.10) will not average to zero. Therefore, the observed frequency of the emitted photon will be

$$f' = f\left(1 + \frac{\langle v^2\rangle}{c^2}\right) \tag{4.11}$$

where $\langle v^2\rangle$ is the average squared vibrational velocity. The frequency shift in equation (4.11) introduces an energy shift, the second order Doppler shift, ΔE_{SOD}, given by

$$\Delta E_{\mathrm{SOD}} = -E_\gamma\left(\frac{\langle v^2\rangle}{2c^2}\right).$$

This corresponds to a velocity shift of the Mössbauer resonance due to the second order Doppler effect of

$$\Delta v_{\mathrm{SOD}} = -\frac{\langle v^2\rangle}{2c}. \tag{4.12}$$

The vibrational velocity in equation (4.12) depends on temperature and also on the properties of the host material, most noticeably on the atomic mass and the Debye temperature of the solid. In principle, the measured center shift of the Mössbauer spectrum is the sum of the isomer shift, as described above, and the second order Doppler shift. In practice, if the source and absorber are at the same temperature and the Debye temperatures of the source matrix and the absorber material are similar, then the effects of the second order Doppler shift will cancel out and it is appropriate to refer to the measured center shift as the isomer shift. This is most commonly the case for Mössbauer results reported in the literature where the effects of the second order Doppler shift are small compared with other uncertainties in the measurements. However, in cases where the temperatures of the source and absorber are substantially different or in cases where the source and absorber materials have substantially different Debye temperatures, then the effects of the second order Doppler shift may become important. In such cases, depending on the information that is to be obtained from the measurement, it may be necessary to make corrections to the measured center shifts.

Bibliography and suggestions for further reading

Chemtob M, Moniz E J and Rho M 1974 Deuteron electromagnetic structure at large momentum transfer *Phys. Rev.* C **10** 344–52

Greenwood N N and Gibb T C 1971 *Mössbauer Spectroscopy* (London: Chapman and Hall)

Onley D S 1968 Corrections to high-energy electron-nucleus scattering *Nucl. Phys.* A **118** 436–48

Shenoy G K and Wagner F E 1978 *Mössbauer Isomer Shifts* (Amsterdam: North-Holland)

Sick I, Bellicard J B, Bernheim M, Frois B, Huet M, Leconte P, Mougey J, Phan X H, Royer D and Turck S 1975 Shell structure of the ^{58}Ni charge density *Phys. Rev. Lett.* **35** 910–3

Tarbutton R M and Davies K T R 1968 Further studies of Hartree–Fock approximation in finite nuclei *Nucl. Phys.* A **120** 1–24

IOP Publishing

The Mössbauer Effect (Second Edition)

Richard A Dunlap

Chapter 5

Hyperfine interactions—part II: the electric quadrupole interaction

5.1 Introduction

The ability of Mössbauer effect experiments to observe the chemical isomer shift is a very valuable feature of this method. However, as indicated in chapter 3, other electromagnetic moments of the nucleus interact with external fields, and these interactions provide a means of studying additional properties of materials. In the present chapter we consider the way in which the electric quadrupole moment of the nucleus can provide useful information through Mössbauer effect measurements.

5.2 The electric quadrupole interaction

Returning to the discussion in chapter 3 concerning electric quadrupole interactions, we look at the third term in equation (3.5) in more detail. This term gives the electric quadrupole interaction energy as

$$U = \frac{1}{2}\sum_{i,j}\frac{\partial^2\phi}{\partial x_i \partial x_j}\int \rho(\boldsymbol{r})x_i x_j d^3\boldsymbol{r} \tag{5.1}$$

where the integral on the right-hand side represents the electric quadrupole moment of the nucleus. For axially symmetric nuclear charge distributions, the quadrupole moment may be written in terms of equation (3.9) as

$$Q = \int \rho(\boldsymbol{r})[3z^2 - r^2]d^3\boldsymbol{r} = \int \rho(\boldsymbol{r})[2z^2 - (x^2 + y^2)]d^3\boldsymbol{r}. \tag{5.2}$$

For ellipsoidal nuclear charge distributions as shown in figure 5.1, equation (5.2) gives the sign of the quadrupole moment as shown in the figure. It is also seen that if the charge distribution of the nucleus is spherically symmetric, then the x-, y- and

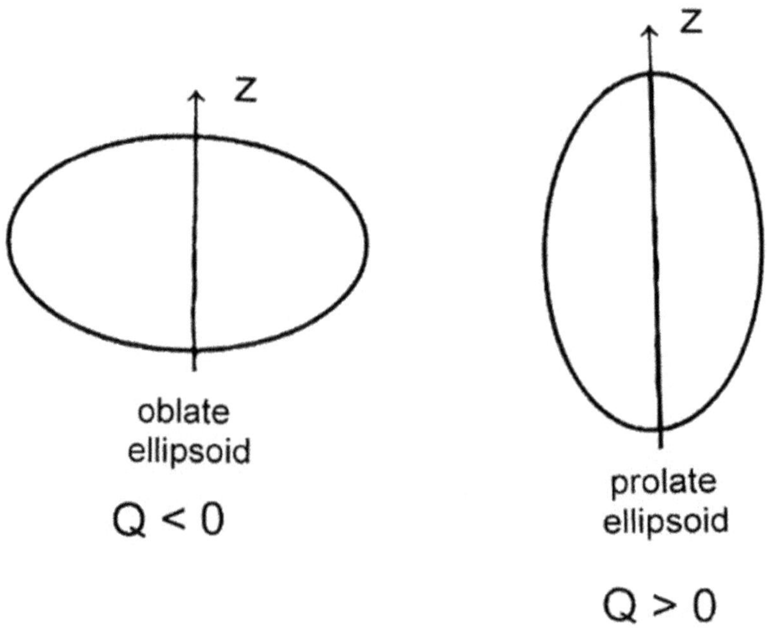

Figure 5.1. Oblate and prolate ellipsoidal nuclear charge distribution giving rise to negative and positive electric quadrupole moments, respectively.

z-directions become equivalent and the integrand, $[2z^2-(x^2+y^2)]$ vanishes giving $Q = 0$.

We now look more carefully at the derivative term in equation (5.1). This is a second derivative of an electrostatic potential. We can understand the physical meaning of this in the following way. The first derivative of an electrostatic potential, i.e., $\partial\phi/\partial x$, is the component of the electric field in the x-direction. The second derivative of the electrostatic potential will be the electric field gradient. The electric field gradient is a tensor with elements as

$$V_{ij} = \frac{\partial^2\phi}{\partial x_i \partial x_j}$$

where $x_i, x_j = x, y, z$.

For nuclear charge distributions with axial symmetry along the z-direction, the maximum electric quadrupole interaction will occur for the z-component of the gradient of the electric field in the z-direction, that is, $\partial^2\phi/\partial z^2$. This is defined as the principal component of the electric field gradient,

$$V_{zz} = \frac{\partial^2\phi}{\partial z^2}.$$

The Laplace equation places the constraint on the diagonal terms of the electric field gradient tensor,

$$\frac{\partial^2 \phi}{\partial x^2} + \frac{\partial^2 \phi}{\partial y^2} + \frac{\partial^2 \phi}{\partial z^2} = 0.$$

By definition, the diagonal components of the electric field gradient follow the relationship

$$|V_{zz}| > |V_{yy}| \geqslant |V_{xx}|$$

and this allows for the definition of the asymmetry parameter, η, as

$$\eta = \frac{V_{xx} - V_{yy}}{V_{zz}}. \tag{5.3}$$

Based on the above relations, η can take on values $0 \leqslant \eta \leqslant 1$.

Using the definitions given above, equation (5.1) may be written as

$$U = e^2 qQ \tag{5.4}$$

where it is customary to define

$$q = \frac{V_{zz}}{e}. \tag{5.5}$$

This is a classical approach to calculating the electric quadrupole interaction energy. A proper quantum mechanical approach gives a similar result of the form

$$U = e^2 qQ \cdot \frac{1}{4}\left[\frac{3{m_I}^2 - I(I+1)}{3I^2 - I(I+1)}\right] \tag{5.6}$$

where I is the total nuclear spin and m_I is the z-component of the total nuclear spin. Each nuclear level has an electric quadrupole interaction that involves the electric quadrupole moment of the nuclear state, Q, the principal component of the electric field gradient, V_{zz}, and the relevant quantum numbers of the nuclear state. The first and third of these factors are properties of the specific nuclear state and will be the same for all Mössbauer experiments involving a particular nuclear level. The second term, V_{zz}, is a function of the electric potential external to the nucleus and will provide information about the electronic environment in which the atom exists.

The sign of the electric quadrupole interaction, as defined in equation (5.6), is determined by the sign of both q (and hence V_{zz}) and Q. The sign of Q, as noted above, is determined by the fundamental properties of the Mössbauer nucleus. Thus, the sign of V_{zz} is determined by the sign of the quadrupole interaction energy as given in equation (5.4). The measurement of the sign of e^2qQ by means of the Mössbauer effect is discussed further in section 5.4. The sign of V_{zz} provides some insight into the origins of the quadrupole interactions. For an ionic crystal, q from equation (5.5) can be written in terms of the Sternheimer shielding-antishielding factors as (Sen 1980),

$$q = \frac{V_{zz}}{e} = (1 - R)q_{\text{val}} + (1 - \gamma_\infty)q_{\text{lat}}$$

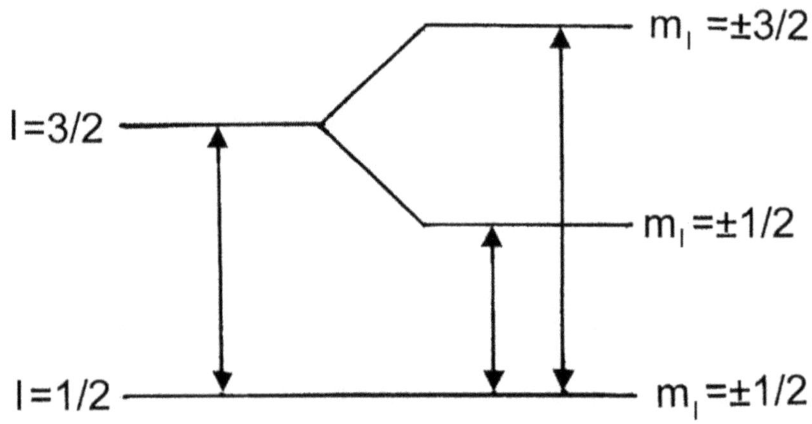

Figure 5.2. Electric quadrupole splitting of the excited state energy level in ^{57}Fe and the two possible transition energies to the ground state.

where R and γ_∞ are factors related to Sternheimer shielding and antishielding, respectively. The first term represents the effect of quadrupole core polarization at the location of the valence electrons and the second term represents the effect of core polarization at a distant point in the lattice. The Sternheimer shielding factor typically takes on values of $0 < R < 1$, while the antishielding factor can take on both positive and negative values and can, in principle, have values as large as $\gamma_\infty \sim 100$. It is clear, therefore, that the sign of the principal component of the electric field gradient, q, depends critically on the relative importance of shielding and anti-shielding and the measurement of the sign of the electric quadrupole interaction provides information on the mechanisms that give rise to this interaction.

In order to understand the implications of equation (5.6), it is interesting to look at a specific example, that is, the 14.4 keV to ground state Mössbauer transition for ^{57}Fe as shown in figure 5.2. This is an $I = 3/2 \rightarrow I = 1/2$ transition. Beginning with the $I = 1/2$ ground state for ^{57}Fe, we see that the allowed values of m_I will be $-1/2$ and $+1/2$. We can use equation (5.6) to calculate the interaction energy for these two values of m_I. It is easy to show that

$$U(m_I = -1/2) = 0 \text{ and } U(m_I = +1/2) = 0.$$

This result provides some fundamental insight into the quantum mechanical properties of nuclei, that is, nuclei with total spin 1/2 are not subject to electric quadrupole interactions because they are spherically symmetric. It is also clear from the form of equation (5.6), that the sign of m_I is not relevant.

If we repeat the above analysis for the $I = 3/2$ excited state of ^{57}Fe, we find that the allowed m_I values are $-3/2$, $-1/2$, $+1/2$ and $+3/2$. A simple analysis shows that

$$U(m_I = \pm 1/2) = -e^2qQ/4$$

and

$$U(m_I = \pm 3/2) = +e^2qQ/4.$$

5.3 Quadrupole splitting of Mössbauer spectra

The result presented in the previous section means that the electric quadrupole interaction has lifted the degeneracy in m_I and split the $I = 3/2$ excited state into two substates, one with energy increased by $eQV_{zz}/4$ and one with energy decreased by $eQV_{zz}/4$. This situation is illustrated in figure 5.2. The figure shows that there are now two possible transitions from the split excited state to the unsplit ground state with different energies. This gives rise to two resonant peaks in the Mössbauer effect spectrum, as illustrated in figure 5.3, where the peaks are separated by an energy

$$\Delta E = \left[\frac{e^2qQ}{4}\right] - \left[-\frac{e^2qQ}{4}\right] = \frac{e^2qQ}{2}.$$

In the Mössbauer spectrum this splitting is observed as a Doppler velocity given by

$$\Delta v = \frac{e^2qQc}{2E_\gamma}.$$

It is important to note that the energy levels in figure 5.2 are not to scale. The electric quadrupole splitting is many orders of magnitude smaller than the separation

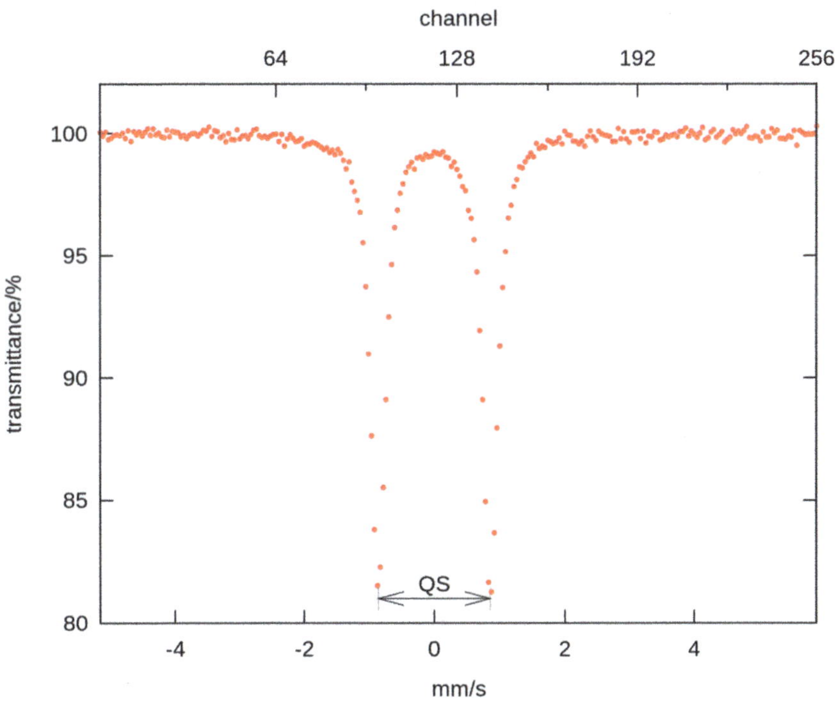

Figure 5.3. The quadrupole splitting in the figure (QS) is given by equation (5.5). This [Room temperature ^{57}Fe Mössbauer effect spectrum of sodium nitroprusside] image has been obtained by the author from the Wikimedia website http://commons.wikimedia.org/wiki/File:Sodium_nitroprusside_Mössbauer_spectrum.svg (Divide 2010), where it is stated to have been released into the public domain. It is included within this chapter on that basis.

between the ground and excited states. It is the hyperfine nature of the quadrupole splitting that allows it to be observed by Doppler shifting the source energy.

For a nuclear transition with a known electric dipole moment, Q, the measurement of the Mössbauer spectra, as in figure 5.3, will allow for the determination of the principal component of the electric field gradient, as in equation (5.6). The electric field gradient is the result of asymmetry in the electric potential and, hence, the electric charge distribution in the environment around the nucleus. This reflects the crystallographic symmetry surrounding the site occupied by the Mössbauer probe nucleus. Therefore, the observed electric quadrupole splitting in Mössbauer spectra is characteristic of the details of the crystal structure, as will be discussed further in chapters 9 and 10.

5.4 Intensity of absorption lines

The two peaks in a quadrupole split ^{57}Fe Mössbauer effect spectrum, as seen in figure 5.3, are the result of the two transitions between the excited state and the ground state, as illustrated in figure 5.2. The relative intensity (i.e., area) of the two peaks in the spectrum is the result of the relative transition probabilities. In general, the transition probability is the product of two components, an angle-independent term and an angle-dependent term. The angle-independent term is by the square of the Clebsch–Gordon coefficient, C^2, for the transition. For an oriented absorber (i.e., single crystal) the angle-dependent term, Θ, is a function of the angle, θ, between the principal electric field gradient direction and the direction of the γ-ray. Referring to the transitions shown in figure 5.2, the angle-dependence of the transition means that the intensity of the Mössbauer peak from the $\pm 3/2 \rightarrow \pm 1/2$ transition can be different than the intensity from the $\pm 1/2 \rightarrow \pm 1/2$ transition and this will lead to an asymmetry of the Mössbauer spectrum. Table 5.1 gives the relative intensities for the two transitions of a quadrupole split $3/2 \rightarrow 1/2$ transition as would be appropriate for ^{57}Fe or ^{119}Sn Mossbauer spectroscopy. Extensive tables of the relative line intensities for quadrupole split spectra for Mössbauer nuclides with spin states other than $3/2 \rightarrow 1/2$ can be found in Greenwood and Gibb (1971).

Polycrystalline samples, i.e., powdered samples or thin foils in which the crystallites are randomly oriented, have lines with relative intensities given by the integral over all angles of the quantities in table 5.1, i.e., 1:1. Figure 5.4 shows a stick

Table 5.1. Angle dependence of the relative intensity of the Mössbauer peaks for the $3/2 \rightarrow 1/2$ transition with electric quadrupole interactions. The total transition probability is given by the product $C^2\Theta$.

Transition (m_I states)	C^2	Θ
$\pm 1/2 \rightarrow \pm 1/2$	1	$2 + 3\sin^2\theta$
$\pm 3/2 \rightarrow \pm 1/2$	1	$3(1 + \cos^2\theta)$

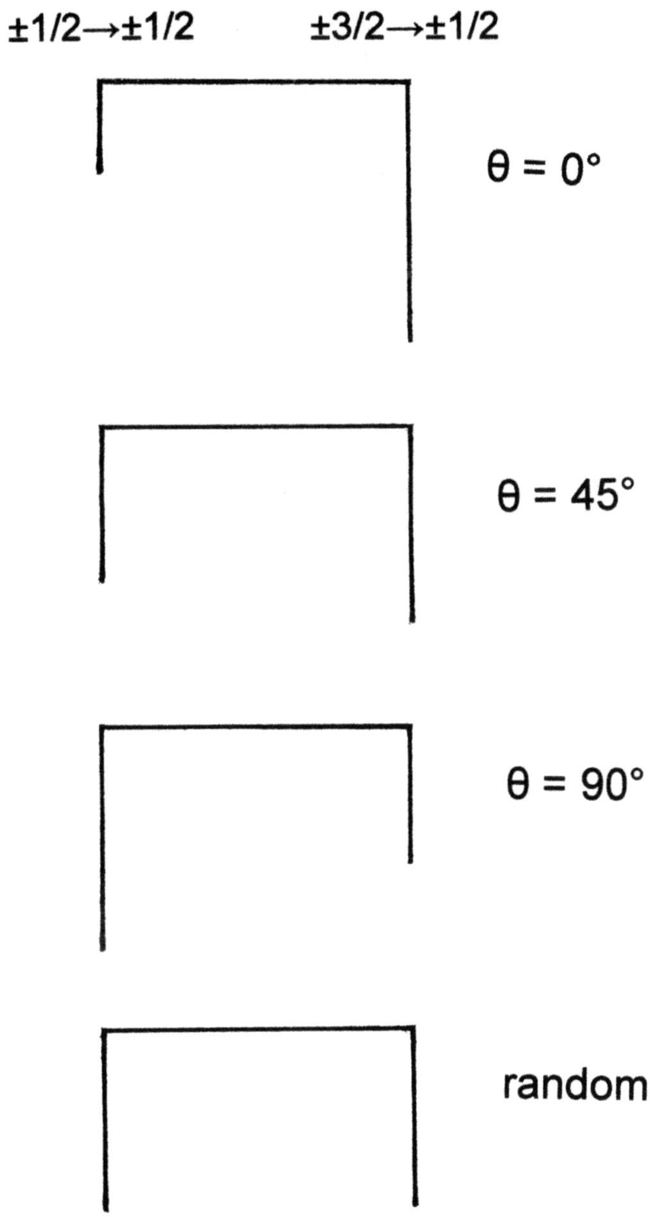

Figure 5.4. Stick diagram showing relative intensity of the two lines in a quadrupole split doublet as a function of the angle between the principal electric field gradient direction and the γ-ray direction as predicted by the equations in table 5.1.

diagram of the relative intensities (areas) of the two Mössbauer peaks as identified by the m_I values of the excited and ground states. The measurement of the quadrupole split spectrum of oriented single crystal absorbers and the determination of the relative intensities of the peaks as illustrated in figure 5.4 allows for both the determination of the direction of the principal electric field gradient component, as well as the sign of the electric field gradient tensor. Neither of these quantities can be determined by measurements on polycrystalline samples.

It should be noted that in some cases, the absorption peaks have different depths but also different widths so as to maintain equal areas. This feature is often characteristic of disordered materials and is discussed in section 10.5.

Bibliography and suggestions for further reading

Divide 2010 *Room Temperature ^{57}Fe Mössbauer Effect Spectrum of Sodium Nitroprusside* (https://commons.wikimedia.org/wiki/File:Sodium_nitroprusside_Mössbauer_spectrum.svg)

Greenwood N N and Gibb T C 1971 *Mössbauer Spectroscopy* (London: Chapman and Hall)

Kalvius M and Kienle P (ed) 2012 *The Rudolf Mössbauer Story: His Scientific Work and Its Impact on Science and History* (Heidelberg: Springer)

Sen K D 1980 Sternheimer quadrupole shielding-antishielding function for F^-, Cl^-, Br^- and I^- in ionic solids *J. Phys. C: Solid State Phys.* **13** L423–6

Wertheim G 1964 *Mössbauer Effect: Principles and Applications* (New York: Academic)

Yoshida Y and Langouche G (ed) 2013 *Mössbauer Spectroscopy: Tutorial Book* (Heidelberg: Springer)

Chapter 6

Magnetic properties of materials

6.1 Introduction

The third, and final, hyperfine interaction that can be observed by means of the Mössbauer effect is the magnetic dipole interaction. Before discussing this interaction in detail, it is important to review the physics of magnetically ordered materials. We begin with a consideration of paramagnetic materials and the reasons for the formation of atomic magnetic moments. Later in the chapter we will summarize the properties of the major types of magnetically ordered materials.

6.2 Paramagnetic materials

Paramagnetic materials consist of atoms which carry magnetic moments. These magnetic moments arise as a result of the angular momenta associated with the particles that comprise the atoms; the electrons, the protons and the neutrons. The contribution from protons and neutrons is very small, as discussed below, so we begin with a detailed consideration of the electrons.

The electrons associated with an atom produce a net magnetic moment, $\boldsymbol{\mu}$, which is related to the overall angular momentum of the electrons, $\boldsymbol{J}$, as

$$\boldsymbol{\mu} = \gamma\hbar\boldsymbol{J} \tag{6.1}$$

where γ is the magnetogyric ratio (sometimes called the gyromagnetic ratio) which relates angular momentum in units of $\hbar$ to the magnetic moment (in SI, the magnetogyric ratio is in units of Joules/Tesla). It should be noted that $\boldsymbol{\mu}$ and $\boldsymbol{J}$ are vector quantities, where the total $\boldsymbol{J}$ is the vector sum of the total spin and orbital components of the angular momentum. Equation (6.1) is often written as

$$\boldsymbol{\mu} = -g\mu_B\boldsymbol{J} \tag{6.2}$$

where g is the dimensionless spectroscopic splitting factor (usually just referred to as the g-factor) and μ_B is the Bohr magneton defined to be

$$\mu_B = \frac{e\hbar}{2m_e}. \tag{6.3}$$

Here e is the electronic charge and m_e is the electron mass. The negative sign in equation (6.2) comes from the fact that the electrons have a negative charge, so the vector magnetic moment will point in the opposite direction to the vector angular momentum (this is easy to see in the classical picture for the orbital component). From equation (6.2) the measured magnetic moment, i.e., the magnitude of the expectation value of the right-hand side of the equation, will be

$$\mu = g\mu_B[J(J+1)]^{1/2}.$$

We could take a similar approach to determining the contribution to the magnetic moment by the neutron and proton. However, while the neutron and proton contribute equally with the electron to the angular momentum in units of $\hbar$, they are much less effective at producing a real magnetic field because of their mass. The nuclear equivalent of the Bohr magneton, the nuclear magneton, would be defined by equation (6.3) with the proton mass (instead of the electron mass) in the denominator, meaning that one unit of angular momentum carried by a proton or neutron produces about 1/2000 of the magnetic moment as it would for the electron and for the present discussion this contribution can be ignored.

For a free electron the g-factor is approximately 2.00. For the electrons in an atom the g-factor, calculated on the basis of the total angular momentum quantum number and the total spin and orbital components, is

$$g = 1 + \frac{J(J+1) + S(S+1) - L(L+1)}{2J(J+1)}.$$

Note that the quantum numbers are represented by upper case letters indicating that they are the quantum numbers for a system of particles (i.e., all of the atomic electrons) and are determined by appropriately summing the components due to the individual electrons.

To determine the magnetic moment associated with an atom it is now a matter of merely determining the total S, L and J for the electrons of the atom. A simple approach might be to consider that all paired electrons have spins and orbital angular momenta that align antiparallel and cancel one another. In practice, this approach is not valid, and we need a more detailed analysis of how the spins align. A set of guidelines, known as the Hund rules, provides some useful insight into the way in which electrons in atoms behave, and also, in many cases, provides a reasonable estimate of the resulting atomic magnetic moment. The Hund rules state that the components of the electron spin for unfilled atomic shells are described in the following way:

1. The total spin, S, is the maximum that is allowed by the Pauli exclusion principle.
2. The orbital angular momentum, L, is the maximum that is consistent with rule 1.

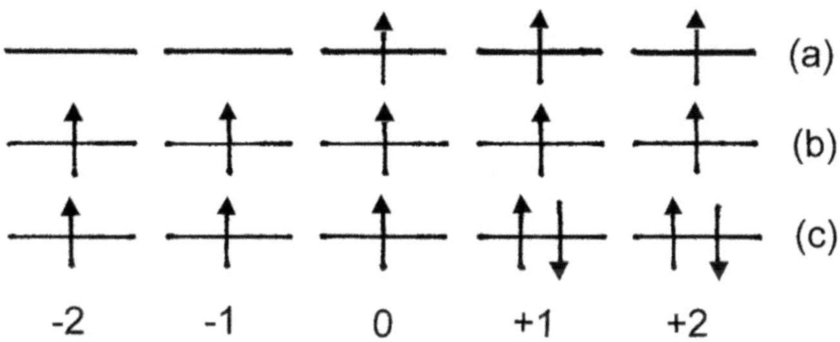

Figure 6.1. Examples of the application of the Hund rules to the 3d shell with (a) 3 electrons, (b) 5 electrons and (c) 7 electrons showing the predicted electron spin states for different m_l values.

3. The total angular momentum, J, is equal to the magnitude of the difference of L and S when the shell is less than half full, and is equal to the magnitude of the sum of L and S when the shell is more than half full. When the shell is exactly half full then rule 1 requires that $L = 0$, so that $J = S$.

An interesting example of the application of the Hund rules considers the 3d transition metals. This example is useful for understanding the properties of many magnetic materials. Here the magnetic moment is the result of an unfilled 3d shell characterized by the quantum numbers $n = 3$ and $l = 2$. The 3d shell is degenerate in m_l, the z-component of l, with $(2l+1)$ possible values of $m_l = -2, -1, 0, +1$ and $+2$. Each of these m_l states is degenerate in the spin component of the angular momentum with $m_s = -1/2$ or $+1/2$. Thus, the total degeneracy of the 3d level will be $2(2l+1) = 2(2\times2+1) = 10$. If the shell is less than or equal to half full then rule 1 requires that each electron will occupy a separate m_l state, each with the same value of m_s. The m_l states which are occupied will be such that rule 2 is satisfied. If the shell is more than half full then some m_l states will have more than one electron (with opposite values of m_s). The states that are occupied with more than one electron are determined by rule 2. Rule 3 tells us about the vector relationship between the spin and orbital angular momenta and how they should be added.

It is useful to look at some examples. We consider atoms with 3, 5 and 7 3d electrons, that is, the 3d shell less than half filled, exactly half filled and more than half filled. Figure 6.1 shows the electron configurations for these cases such that the Hund rules are satisfied. The resulting values of S, L and J are given in table 6.1.

Once the values of S, L and J are determined by the Hund rules, the value of the g-factor can be determined, and the predicted magnetic moment (usually expressed in units of the Bohr magneton) can be determined. We can apply this approach to the neutral atoms of the elements in the 3d series of the periodic table, but it is often more relevant to consider ions of these elements, as would be appropriate for ionic compounds where the atoms are in well defined valence states. Some results for 2^+

Table 6.1. Summary of angular momentum quantum numbers for the example in figure 6.1.

3D electrons	S	L	J
3	3/2	3	3/2
5	5/2	0	5/2
7	3/2	3	9/2

Table 6.2. Calculated and typical measured values of the magnetic moment (in Bohr magnetons) for some 2+ ions of 3d transition metal elements.

Ion	3D electrons	μ/μ_B (calculated)	μ/μ_B (typical)
V^{2+}	3	0.77	3.8
Cr^{2+}	4	0	4.9
Mn^{2+}	5	5.92	5.9
Fe^{2+}	6	6.70	5.4
Co^{2+}	7	6.63	4.8
Ni^{2+}	8	5.59	3.2
Cu^{2+}	9	3.55	1.9

ions of 3d elements are summarized in table 6.2. It is important to note that the 4s electrons also contribute to bonding and are typically delocalized before the 3d electrons. The results illustrated in the table show that the application of the Hund rules successfully predicts the magnetic moment in some cases and shows the correct trend (at least) for the latter half of the 3d series. In any case, the results demonstrate that the prediction of magnetic moments is much more complex than just looking at whether the number of electrons is even or odd and show that a careful consideration of all aspects of the electron angular momentum is necessary.

The orientation of the magnetic moments in a material is influenced by several factors, particularly, the temperature and the interaction between the magnetic moments in the material and any applied magnetic fields. In materials in which the interactions between the magnetic moments are relatively weak, then the material may be described as a paramagnet. The interaction between an external magnetic field and the magnetic moments associated with the atoms in a paramagnetic material is illustrated in figure 6.2. In the absence of an applied magnetic field, as shown in figure 6.2(a), the magnetic moments will point in random dimensions. This is the result of the thermal energy associated with the magnetic moments at finite temperature which causes the direction of each moment to move around at random. Thus, figure 6.2(a) may be considered as a snapshot of the paramagnet at a moment in time. When an external magnetic field is applied to the system, then the

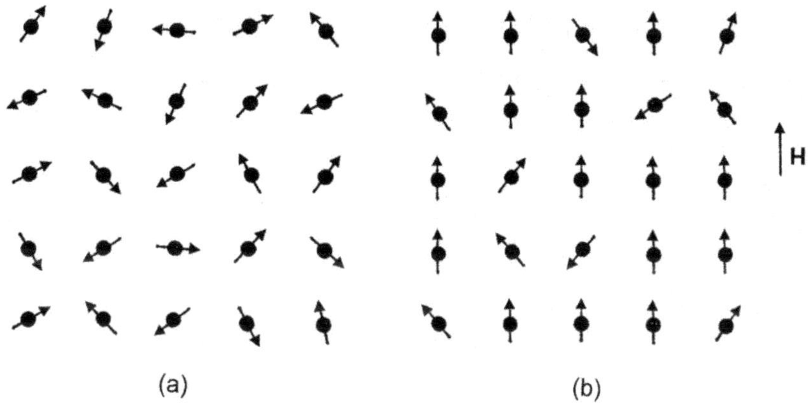

Figure 6.2. Alignment of magnetic moments in a paramagnetic material (a) without an applied magnetic field and (b) with an applied magnetic field.

interaction between the field and the magnetic moments of the atoms causes the moments to preferentially align in a direction parallel to the applied field as shown in figure 6.2(b). In this case, we can define the net magnetization of the sample as the vector sum of the individual atomic magnetic moments,

$$\boldsymbol{M} = \frac{1}{V}\sum \boldsymbol{\mu}_i \tag{6.4}$$

Here the sum of the vector moments is divided by the volume of the sample to yield a magnetization per unit volume so that the size of the sample is not relevant, and the result is an intrinsic property of the particular material. It is easy to see that this vector sum for the paramagnet without an applied magnetic field, as shown in figure 6.2(a), will be zero, while it will be non-zero for the sample in an applied magnetic field, as in figure 6.2(b). One can infer from this behavior that the net magnetization will increase with increasing applied magnetic field, H. This is expressed as

$$\boldsymbol{M} = \chi_V \boldsymbol{H}$$

where the proportionality constant, χ_V, is the volumetric magnetic susceptibility.

In figure 6.2(b), the temperature also plays a role. As the temperature increases, it becomes more difficult for the applied magnetic field to align the magnetic moments as the thermal energy tends to misalign them. The temperature dependence of the paramagnetic susceptibility is given by the Curie law,

$$\chi_V = \frac{C}{T}. \tag{6.5}$$

The Curie law is based on a simple two-state model and is quite effective at predicting the magnetic behavior of many paramagnetic materials. Figure 6.3 shows some experimental results for the temperature dependence of the paramagnetic

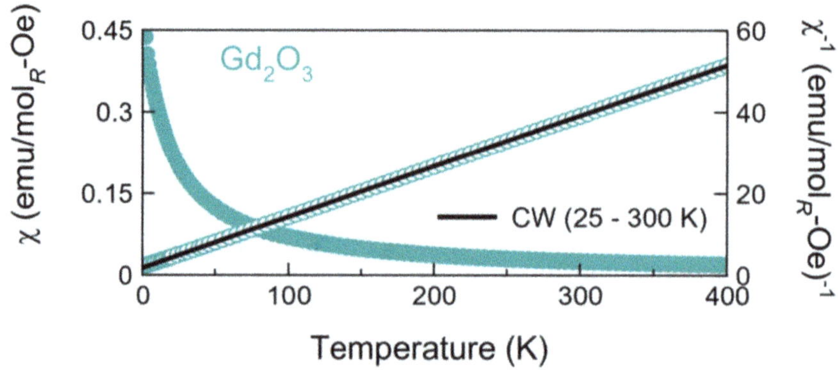

Figure 6.3. Temperature dependence of the magnetic susceptibility of Gd_2O_3. The plot also shows the susceptibility (solid data points, left scale) and the linear temperature dependence of the inverse susceptibility as given by equation (6.5) (open data points, right scale). CC BY 4.0. Reprinted by permission from Mugiraneza and Hallas (2022) CC BY 4.0.

susceptibility of Gd_2O_3. The validity of the Curie law is demonstrated by plotting the inverse of the susceptibility as a function of temperature. The Curie law predicts a linear relationship between $1/\chi_V$ and T with an intercept at $T = 0$. The figure shows that this behavior is observed over a large range of temperatures.

6.3 Ferromagnetic materials and mean field theory

Ferromagnets are materials in which there is a preferential parallel alignment of the magnetic moments below a particular temperature, the Curie temperature, T_c. This preferential alignment of the moments results in a net spontaneous magnetization below T_c. The moment alignment for an ideal two-dimensional ferromagnet with a square lattice at zero temperature is illustrated in figure 6.4. The simplest approach to modelling the behavior of a ferromagnet is mean field theory (sometimes called molecular field theory). Mean field theory assumes that the interactions between the magnetic moments may be described by an exchange field, $\boldsymbol{H}_E$, which is proportional to the magnetization,

$$\boldsymbol{H}_E = \lambda \boldsymbol{M} \tag{6.6}$$

where the constant λ is the exchange coupling constant and is specific to the particular material. The exchange field is a vector which points in the direction of the magnetization and tends to orient the magnetic moments along that same direction. Thermal energy tends to randomize the moment orientation and at sufficiently high temperatures (above the Curie temperature) will destroy the ferromagnetic alignment and the material will become a paramagnet.

Above the Curie temperature the magnetic behavior of a material may be modelled by assuming that the magnetization is the sum of the components due to an applied field, H, and the exchange field, H_E,

$$M = \chi_V (H + H_E) \tag{6.7}$$

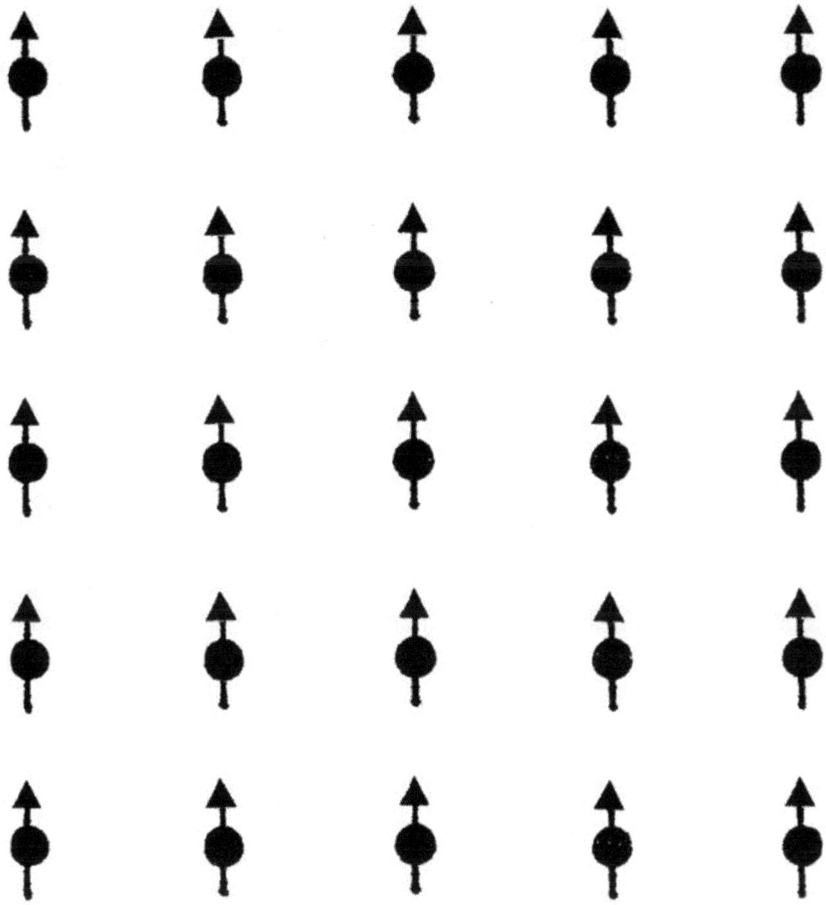

Figure 6.4. Magnetic moment orientation for an ideal two-dimensional ferromagnet at zero temperature.

If we substitute equation (6.6) into (6.7) and use the paramagnetic susceptibility as defined by equation (6.5), $\chi_V = C/T$, we obtain

$$MT = C(H + \lambda M). \tag{6.8}$$

The susceptibility above T_c may, therefore, be written as

$$\chi_V = \frac{M}{H} = \frac{C}{T - T_c} \tag{6.9}$$

where C is the Curie constant and the Curie temperature (sometimes called the critical temperature), T_c, is given in terms of the exchange coupling constant as

$$T_c = C\lambda. \tag{6.10}$$

Therefore, mean field theory shows that the paramagnetic susceptibility diverges as a function of temperature, not above absolute zero, as in the case of an ideal paramagnet, but above the Curie temperature. The mean field result from equation (6.9) is referred to as the Curie–Weiss law.

The behavior below the Curie temperature is the most relevant to the following discussion on hyperfine interactions and may be viewed in the context of mean field theory in the following way. If M is large, then the relationship between the exchange field and the magnetization may be nonlinear and we need to keep the next highest order term in the relationship in equation (6.6),

$$H_E = \lambda(M - \lambda' M^3).$$

Equation (6.8) may then be rewritten as

$$MT = C(H + \lambda M - \lambda' M^3). \tag{6.11}$$

Below the Curie temperature the spontaneous magnetization refers to the $H = 0$ situation (i.e., no applied field) so equation (6.11) can be solved for M as

$$M = \frac{1}{\sqrt{C\lambda'}}[T_c - T]^{1/2} \tag{6.12}$$

where we have used the definition of the Curie temperature as in equation (6.10).

The temperature dependence of the susceptibility above T_c and the spontaneous magnetization below T_c are fundamental properties which describe the behavior of a ferromagnet and equations (6.9) and (6.12) show the theoretical results based on the mean field approximation. In the paramagnetic regime, the susceptibility diverges as the Curie temperature is approached from above. This behavior results from the increasing ease of aligning the magnetic moments with an externally applied magnetic field due to a combination of decreasing thermal energy and increasing magnetic coupling. In the ferromagnetic regime, the spontaneous magnetization increases with decreasing temperature as the thermal excitations, which tend to randomize the moment orientations, decrease and the magnetic coupling becomes relatively more important.

Mean field theory provides some insight into magnetic behavior of ferromagnetic materials and shows reasonable agreement with experimental results for many materials. Figure 6.5 shows a comparison of the mean field result and the measured spontaneous magnetization for nickel. More detailed models, which include information about lattice structure, dimensionality and details of the interactions between moments, provide improved agreement with experimental results for specific materials.

6.4 Antiferromagnetic materials

In an antiferromagnet, an ordered state occurs below a critical temperature, called the Néel temperature, T_N, after Louis Néel (1904–2000), where magnetic moments align antiparallel with their neighboring moments, as illustrated in figure 6.6. The net magnetization will be zero in this ordered state, as equation (6.4) shows that the vector sum of antiparallel moments will be zero.

The paramagnetic susceptibility above the Néel temperature may be expressed as

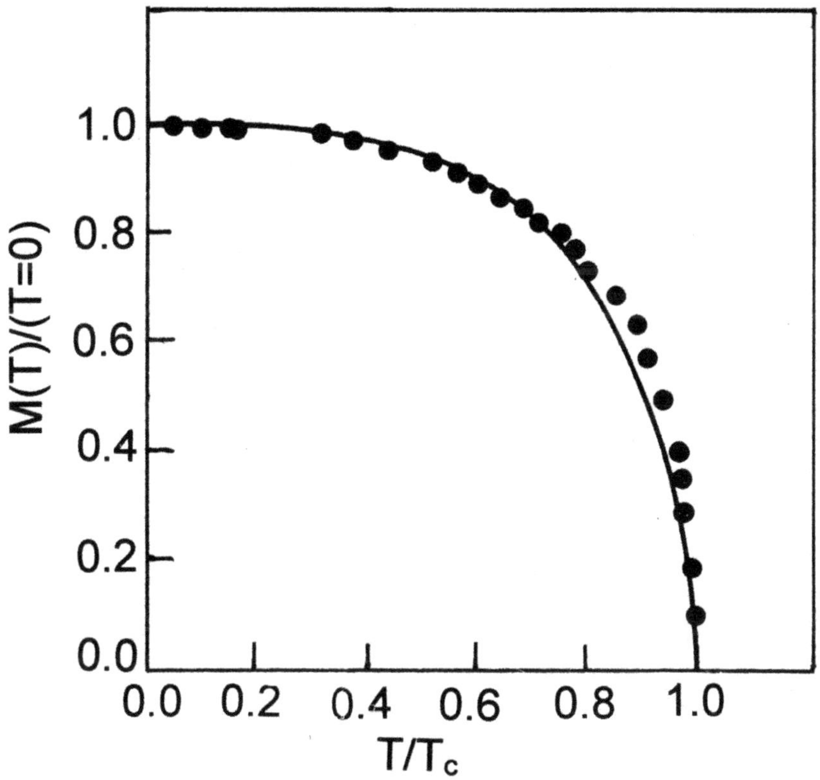

Figure 6.5. Spontaneous magnetization of Ni as a function of temperature below the Curie temperature. The mean field approximation is shown by the solid line. Experimental data are from Weiss and Forrer (1926).

$$\chi_V = \frac{C}{T + T_N}.$$

This behavior shows a cusp (not a divergence) of the susceptibility as the critical temperature is approached from above. Below the Néel temperature, the staggered magnetization is defined as the vector sum of the magnetic moments on a sublattice where the moments point in the same direction. The staggered magnetization shows a temperature dependence which follows along the lines of the spontaneous magnetization of a ferromagnet. This description of the antiferromagnet is relevant to the understanding of Mössbauer effect studies of antiferromagnetic materials, as it is the local environment around the probe site, and not the bulk properties of the material, that is relevant to the hyperfine interactions that influence the Mössbauer spectrum.

6.5 Ferrimagnetic materials

At low temperature, ferrimagnets show the magnetic moment arrangement illustrated in figure 6.7. A ferrimagnet is similar to an antiferromagnet except that the

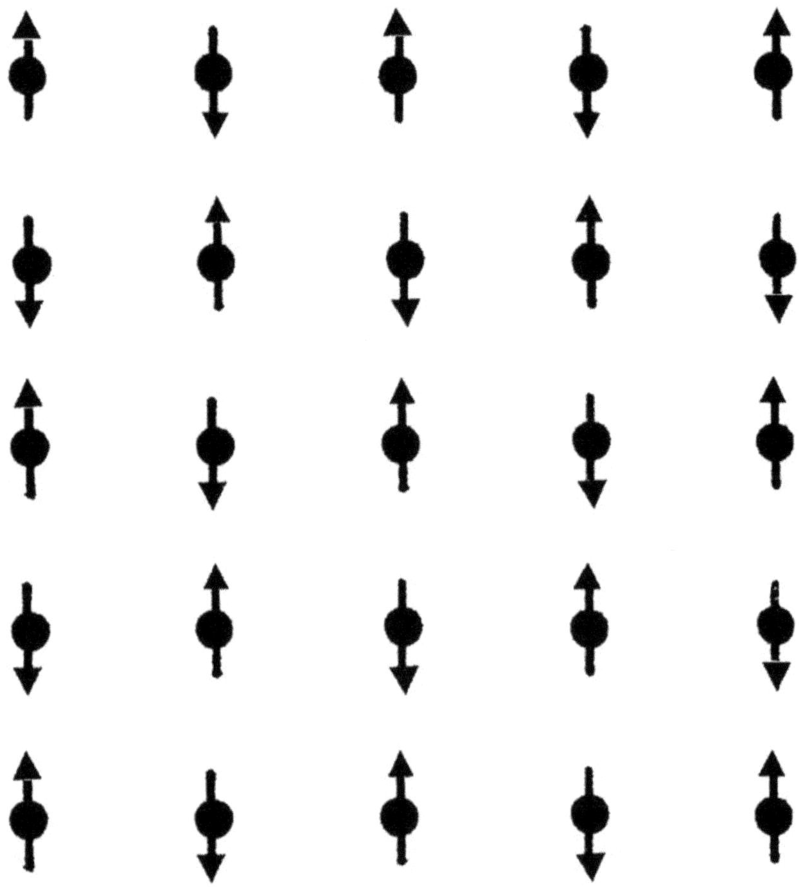

Figure 6.6. Magnetic moment orientation for an ideal two-dimensional antiferromagnet at zero temperature.

magnetic moments associated with one sublattice have a different magnitude than those associated with the other sublattice, so that the vector sum of the moments does not cancel. This leads to a spontaneous magnetization similar to the ferromagnet.

One well known example of a ferrimagnet is Fe_3O_4 (known as magnetite). Chemically the formula for magnetite may be written as $Fe^{2+}O{\bullet}Fe^{3+}{}_2O_3$, where each formula unit contains two iron ions in the 3+ state and one iron ion in the 2+ state. We would expect that iron in different valence states would carry different magnetic moments. Following from the discussion of the Hund rules earlier in the present chapter, we can consider the possible moments associated with these different iron ions. Table 6.3 gives the results of calculations based on the Hund rules for the free Fe^{2+} and Fe^{3+} ions. Since the 3+ ion has a 3d level that is exactly half filled, then the Hund rules state that the orbital angular momentum will be quenched, and the net spin will be due entirely to the vector sum of the intrinsic spins. It is sometimes found that 3d levels that are not exactly half filled can also have an electronic configuration

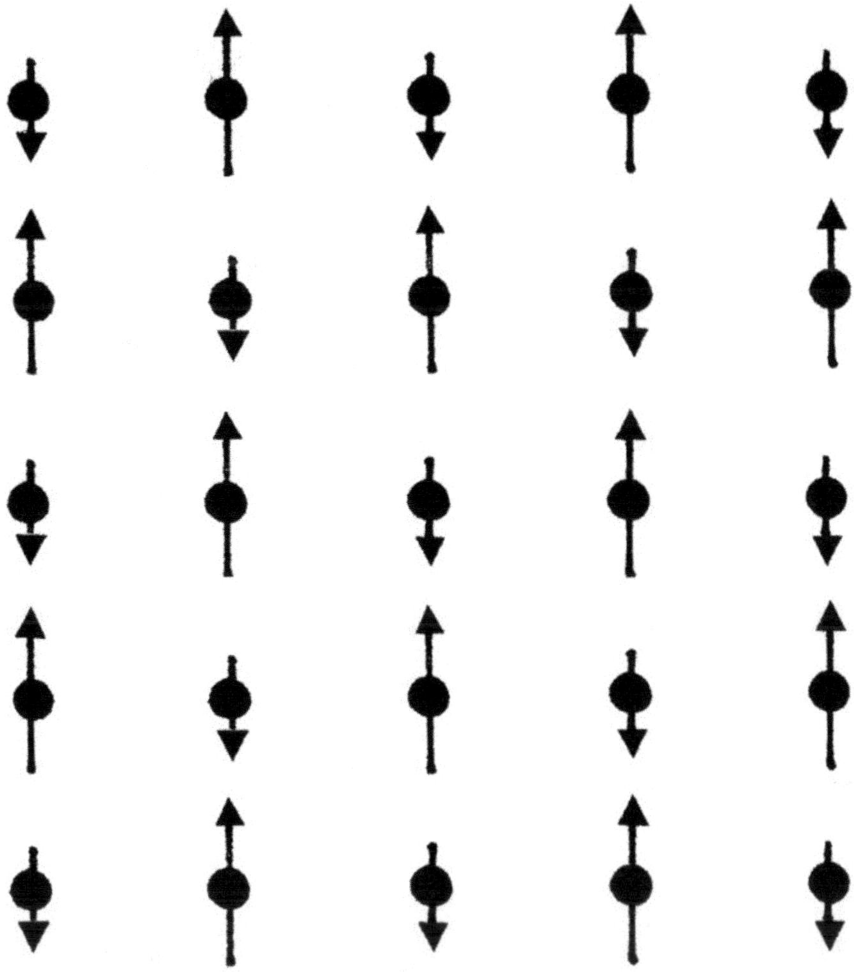

Figure 6.7. Magnetic moment orientation for a simple ideal two-dimensional ferrimagnet at zero temperature.

Table 6.3. Calculated free ion magnetic moments for iron in 2+ and 3+ valence states. Note: [Ar] = $1s^2 2s^2 2p^6 3s^2 3p^6$.

Valence state	Electron configuration	μ/μ_B Hund rules	μ/μ_B L quenched
2+	[Ar]$3d^6$	6.70	4.90
3+	[Ar]$3d^5$	5.92	5.92

where the orbital moment is quenched. The table shows this result for the 2+ ion. If the iron moments all align parallel, then we would expect a total moment per formula unit of $2 \times 5.92 + 6.70 = 18.54\ \mu_B$ if the orbital moment is not quenched and $2 \times 5.92 + 4.90 = 16.74\ \mu_B$ if the orbital moment is quenched. Experimental

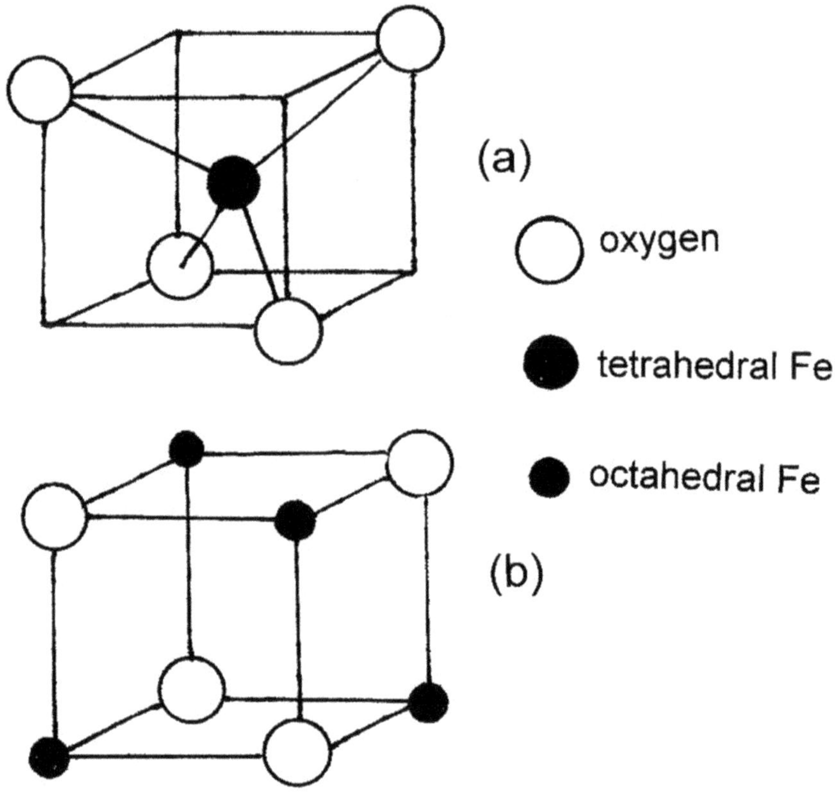

Figure 6.8. Iron sites in the crystal structure of Fe_3O_4; (a) tetrahedral (A-sites) and (b) octahedral (B-sites).

measurements show that the actual magnet moment per formula unit is 4.1 μ_B. In order to understand how this result can be explained, we need to look in detail at the crystal structure of Fe_3O_4.

The basic structure of Fe_3O_4, commonly known as magnetite, is the spinel structure, which describes a wide variety of compounds with the chemical formula AB_2O_4. In the so-called normal spinel structure, the A-atoms occupy tetrahedral sites as shown in figure 6.8(a) and the B-atoms occupy octahedral sites as shown in figure 6.8(b). It is easy to see in the figure that the A-atoms are at the center of a tetrahedron of four oxygen atoms and the B-atoms are at the center of an octahedron of six oxygen atoms if the cubic cell in figure 6.8(b) is repeated. Fe_3O_4 is of the inverse spinel structure where A and B are both iron and the Fe^{2+} ions occupy half of the octahedral sites; Fe^{3+} ions occupy the other half of the octahedral sites and the remaining Fe^{3+} ions occupy all of the tetrahedral sites. Thus, there are three different categories of iron ions in the Fe_3O_4 structure, Fe^{2+} octahedral, Fe^{3+} octahedral and Fe^{3+} tetrahedral, where one formula unit contains one of each type.

In the magnetically ordered ferrimagnetic state of Fe_3O_4, the magnetic moments of the Fe^{2+} octahedral ions align parallel to the magnetic moments of the Fe^{3+} octahedral ions and these both align antiparallel to the magnetic moments of the Fe^{3+}

tetrahedral ions. This means that the octahedral and tetrahedral Fe^{3+} ion moments cancel each other leaving a net magnetization that comes only from the moments on the Fe^{2+} ions. Our simple calculation of the net moment based on quenched orbital momenta would give a total moment per formula unit of 4.9 μ_B, in relatively good agreement with the experimental value of 4.1 μ_B. The effect of this magnetic ordering on the measured Mössbauer effect spectrum will be discussed in chapter 9.

Bibliography and suggestions for further reading

Chen C W 1986 *Magnetism and Metallurgy of Soft Magnetic Materials* (New York: Dover)

Ma S K 1976 *Modern Theory of Critical Phenomena* (Reading, MA: Benjamin/Cummings)

Mugiraneza S and Hallas A M 2022 Tutorial: a beginner's guide to interpreting magnetic susceptibility data with the Curie–Weiss law *Commun. Phys.* **5** 95

O'Handley R C 2000 *Modern Magnetic Materials: Principles and Applications* (New York: Wiley)

Spauldin N 2003 *Magnetic Materials* (Cambridge: Cambridge University Press)

Weiss P and Forrer R 1926 Aimantation et phénomène magnétocalorique du nickel *Ann. Phys.* **10** 153–213

IOP Publishing

The Mössbauer Effect (Second Edition)

Richard A Dunlap

Chapter 7

Hyperfine interactions—part III: the magnetic dipole interaction and the nuclear Zeeman effect

7.1 Introduction

In materials that are magnetically ordered, magnetic fields will exist within the material, and, in particular, at the nuclei of probe atoms used for Mössbauer spectroscopy. These fields may be a local field that results from the magnetic moment associated with the probe atom itself or a transferred field that results from the ordering of magnetic moments associated with neighboring atoms, or both. The magnetic field that exists at the nucleus is referred to as the hyperfine magnetic field, H_{hf}. It is the interaction between the hyperfine field and the magnetic moment (i.e., the dipole moment) of the nucleus that gives rise to changes in the nuclear energy levels that can be observed by the Mössbauer effect as discussed in detail below.

7.2 The magnetic dipole interaction

The interaction energy between the hyperfine field and the magnetic moment of the nucleus is given by

$$U = -g\mu_N \boldsymbol{I} \cdot \boldsymbol{H}_{hf}. \tag{7.1}$$

Here μ_N is the nuclear magneton and, analogous to the Bohr magneton, is defined as

$$\mu_N = \frac{e\hbar}{2m_\mathrm{p}}$$

where m_p is the mass of the proton. In equation (7.1) the nuclear g-factor is defined in terms of the nuclear magnetic moment, μ, as

$$g = \frac{\mu}{\mu_N I}.$$

doi:10.1088/978-0-7503-6039-5ch7

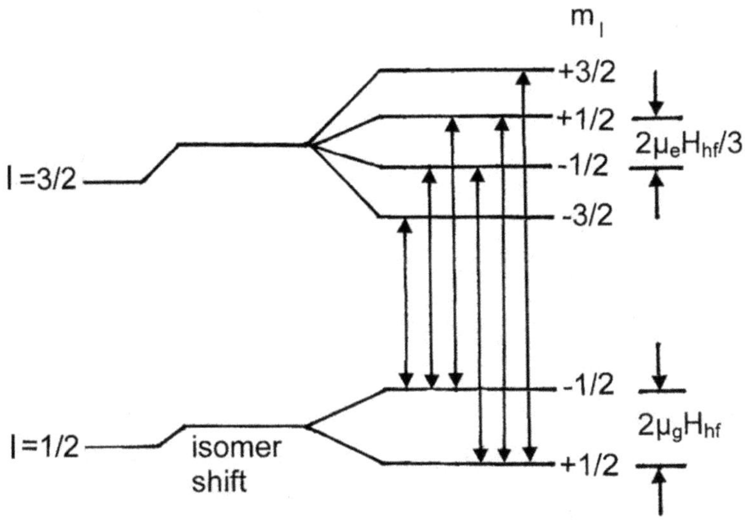

Figure 7.1. Hyperfine splitting of the 3/2 → 1/2 transition in the presence of a hyperfine magnetic field. This diagram is appropriate for ^{57}Fe and the inversion of the signs of the m_I states for the ground and excited states is the result of differences in the sign of the nuclear magnetic moments of these two states. Note: the vertical energy axis is not to scale as the magnetic splitting is many orders of magnitude smaller than the energy difference between the ground and excited states.

It is customary to define the hyperfine magnetic field direction along the *z*-axis, so that equation (7.1) may be written as

$$U = -\frac{\mu m_I H_{hf}}{I}. \tag{7.2}$$

We can see that the energy associated with the interaction of the nuclear magnetic moment and a hyperfine magnetic field depends on several factors: particularly, the magnitude of the magnetic field, the nuclear spin and its *z*-component and the nuclear moment and its sign. It is relevant to look at a particular example. Let us consider a nucleus which has an excited state spin of 3/2 and a ground state spin of 1/2. As well, assume that the magnetic moments of the two states are, $\mu_g > 0$ and $\mu_e < 0$. These conditions are appropriate for the 14.4 keV transition in ^{57}Fe. The excited $I = 3/2$ excited state has four allowed m_I sublevels with $m_I = -3/2, -1/2, +1/2$ and $+3/2$, while the $I = 1/2$ ground state has two allowed m_I states with $m_I = -1/2$ and $+1/2$. The energy shifts for the four excited state sublevels and the two ground state sublevels, as given by equation (7.2), are illustrated in figure 7.1. Note, that in accordance with previous discussions, the isomer shift is always present. This splitting of the nuclear energy levels in the presence of a magnetic field is referred to as the nuclear Zeeman effect.

Figure 7.1 shows the allowed transitions between the excited state and the ground state. While, in principle, there would be eight possible transitions between the four excited sublevels and the two ground sublevels, only six of these are quantum mechanically allowed. The reason for this is the requirement for conservation of angular momentum. During the process of undergoing a transition from the excited state to the ground state, the nucleus emits a gamma ray (photon). The photon has one unit of intrinsic spin ($s = 1$). Thus, depending on the orientation of the spins of the excited state and ground state, the photon can change the z-component of the total angular momentum of the nucleus by −1, 0 or +1 (as measured in units of $\hbar$). This leads to the quantum mechanical selection rule $\Delta m_J = -1,\ 0,\ +1$ and the allowed transitions as shown in the figure. The same selection rules apply to the absorption of a photon which excites the nucleus from the ground state to the excited state.

7.3 Zeeman splitting of Mössbauer spectra

The Mössbauer spectrum for a material which shows magnetic order, and hence Zeeman splitting, is illustrated in figure 7.2. In this case, the spectrum is for elemental α-Fe which is ferromagnetic at room temperature, and which shows a sextet spectrum as a result of the six allowed transitions described above. The hyperfine field is largely due to the local iron magnetic moment on the probe atom. The symmetry of the spectrum about the center position, determined by the isomer shift of iron relative to the host environment for the source, can be explained as follows. The splitting between lines 1 and 2 (counting from the left side of the spectrum) will be the same as the splitting between lines 2 and 3, between lines 4 and 5 and between lines 5 and 6, and the Doppler velocity associated with this splitting is found from equation (7.2) to be

$$v_{1,\,2} = v_{2,\,3} = v_{3,\,4} = v_{4,\,5} = \frac{2\mu_e H_{hf} c}{3E_\gamma}. \tag{7.3}$$

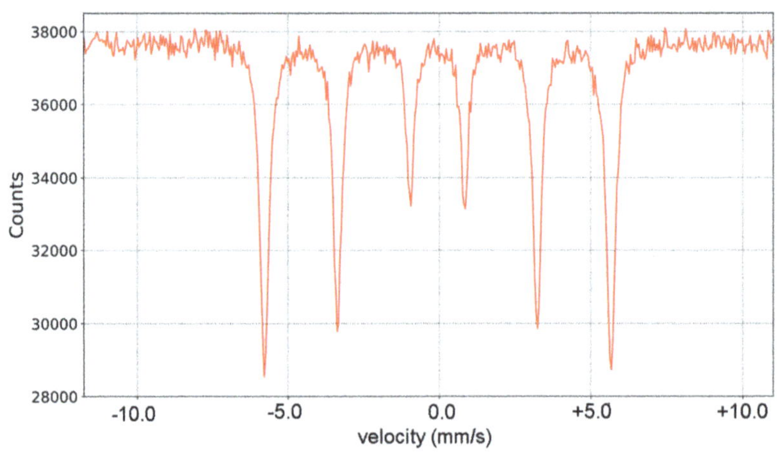

Figure 7.2. Magnetic Zeeman splitting of the Mössbauer effect spectrum of iron (α-Fe) at room temperature. This [Mössbauer spectrum of ^{57}Fe] image has been obtained by the author from the Wikimedia website where it was made available by The High Fin Sperm Whale (2017) under a CC BY-SA 4.0 licence. It is included within this chapter on that basis. It is attributed to The High Fin Sperm Whale.

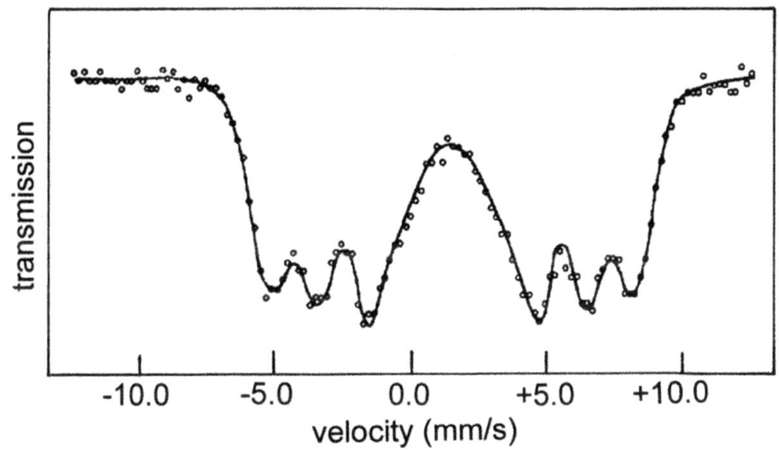

Figure 7.3. Room temperature ^{119}Sn Mössbauer spectrum of the Heusler alloy Co_2MnSn. Adapted from Dunlap and Jones (1982). Copyright 1982 by the American Physical Society.

The splitting between lines 3 and 4 is found to be

$$v_{3,\,4} = \frac{2H_{hf}c}{E_\gamma}\left[\mu_g - \frac{\mu_e}{3}\right]. \tag{7.4}$$

Thus, a knowledge of the nuclear moments will allow for the determination of the magnetic hyperfine field and an *a priori* knowledge of the hyperfine field will allow for the determination of both the ground state and excited state moments.

By comparison to figure 7.2, figure 7.3 shows the Zeeman split ^{119}Sn Mössbauer spectrum of the Heusler alloy Co_2MnSn. In this case, although the material is ferromagnetic due to magnetic moments associated with the Co and Mn atoms, the Sn atoms do not carry a magnetic moment and the magnetic hyperfine field at the Sn nuclei is a transferred field due to the neighboring magnetic moments. ^{119}Sn is also a $3/2 \rightarrow 1/2$ transition and subsequently (like Fe) shows a Zeeman split sextet when the material is magnetically ordered. The difference in the relative splitting of lines 3 and 4 comes from the different ground state and excited state magnetic moments associated with the ^{119}Sn nucleus compared to the ^{57}Fe nucleus.

7.4 Intensity of absorption lines

As discussed in section 5.4 for quadruple split Mössbauer effect spectra, the relative intensities of the six peaks in a Zeeman split spectrum of a magnetic absorber can provide important information about the physical properties of the material. Analogous to this previous discussion, the relative intensities of the six spectral peaks are determined by the relative transition probabilities of the six allowed transitions shown in figure 7.1. These transition probabilities are the product of an angle-independent term, given by the square of the appropriate Clebsch–Gordon coefficient, C^2, and an angle-dependent term, Θ. In the case of a spectrum showing Zeeman splitting by the magnetic dipole interaction, the angle-dependent term is a

Table 7.1. Angle dependence of the relative intensity of the Mössbauer peaks for the 3/2 → 1/2 transition with magnetic dipole interactions. The total transition probability is given by the product $C^2\Theta$.

Transition (m_I states)	C^2	Θ
+3/2 → +1/2	3	$1 + \cos^2\theta$
+1/2 → +1/2	2	$2\sin^2\theta$
−1/2 → +1/2	1	$1 + \cos^2\theta$
+1/2 → −1/2	1	$1 + \cos^2\theta$
−1/2 → −1/2	2	$2\sin^2\theta$
−3/2 → −1/2	3	$1 + \cos^2\theta$

function of the angle, θ, between the magnetization axis and the direction of the γ-ray. Calculated values of C^2 and Θ for the six allowed transitions, as given by their values of m_I, are summarized in table 7.1 for the 3/2 → 1/2 transition appropriate for ^{57}Fe. The stick diagram in figure 7.4 illustrates the relative line intensities for different values of θ. The random case has relative line intensities of 3:2:1:1:2:3. These intensities correspond to an angle of 54.7°, which is referred to as the magic angle. Extensive tables of the relative line intensities for Zeeman split spectra for Mössbauer nuclides with spin states other than 3/2→1/2 can be found in Greenwood and Gibb (1971).

Powdered samples or thin foils in which there is no magnetic texture, have lines with relative intensities given by the integral over all angles of the quantities in table 7.1, i.e., 3:2:1:1:2:3. It is typical for the experimental spectra of powdered samples to have line intensities that are close to these theoretical values. However, it is common for foil samples of ferromagnetic materials to have some preferred domain orientation because of their shape. In such cases, the intensity of the second and fifth peaks in the Mössbauer spectrum will take on values that are different than those shown for the random sample in figure 7.4. The intensity of these peaks can be controlled experimentally through domain orientation by the application of an external magnetic field or, in some cases, through magnetoelastic coupling by the application of uniaxial stress.

7.5 Combined hyperfine interactions

In all cases, the isomer shift is present in addition to electric quadrupole or magnetic dipole interactions. This merely shifts the center position of the doublet or sextet pattern without affecting the spacing or intensity of the peaks. Thus far, in chapters 5 and 7 the features of Mössbauer effect spectra where electric quadrupole interactions or magnetic dipole interactions are present individually, have been discussed. In most cases, for magnetically ordered (e.g., ferromagnetic) materials the electric quadrupole interaction (if present) is small compared to the magnetic dipole interaction and can be ignored. However, in cases where both interactions are present and are significant, the resulting spectra can be quite complex. It is often

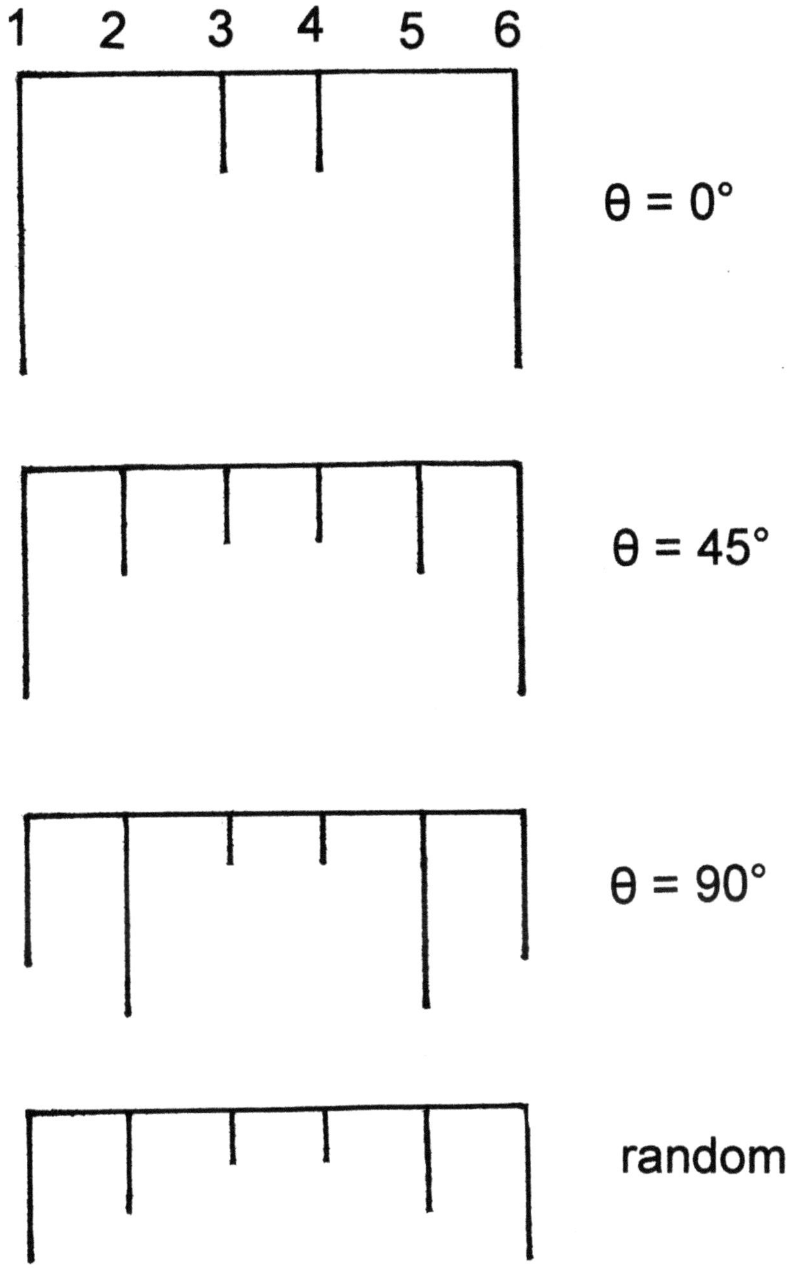

Figure 7.4. Stick diagram showing relative intensity of the six lines in a Zeeman split sextet as a function of the angle between the magnetization direction and the γ-ray direction as predicted by the equations in table 7.1. Values of m_I for the six transitions as labelled in the figure are (1) +3/2 → +1/2, (2) +1/2 → +1/2, (3) −1/2 → +1/2, (4) +1/2 → −1/2, (5) −1/2 → −1/2 and (6) −3/2 → −1/2.

suitable to make the approximation that the electric quadrupole interaction is much weaker than the magnetic dipole interactions (as is typically the case). That is, comparing equations (5.3) and (7.2) gives the relationship

$$eQV_{zz} \ll \mu H_{hf}.$$

In this case, the energy eigenvalues of the nuclear states can be found to be

$$E = -\frac{\mu m_I H_{hf}}{I} + (-1)^{|m_I|+1/2}\frac{eqV_{zz}}{4}\left(\frac{3\cos^2\theta - 1}{2}\right),$$

where θ is the angle between the principal electric field gradient direction and the magnetic axis. For the 3/2 → 1/2 transition, the energy levels are shown in figure 7.5.

The effect of the quadrupole perturbation to a Zeeman split spectrum is to displace the four inner lines of the sextet relative to the two outer lines of the sextet. This effect is known as the quadrupole shift and is illustrated for the spectrum of FeF_2 at low temperatures in figure 7.6.

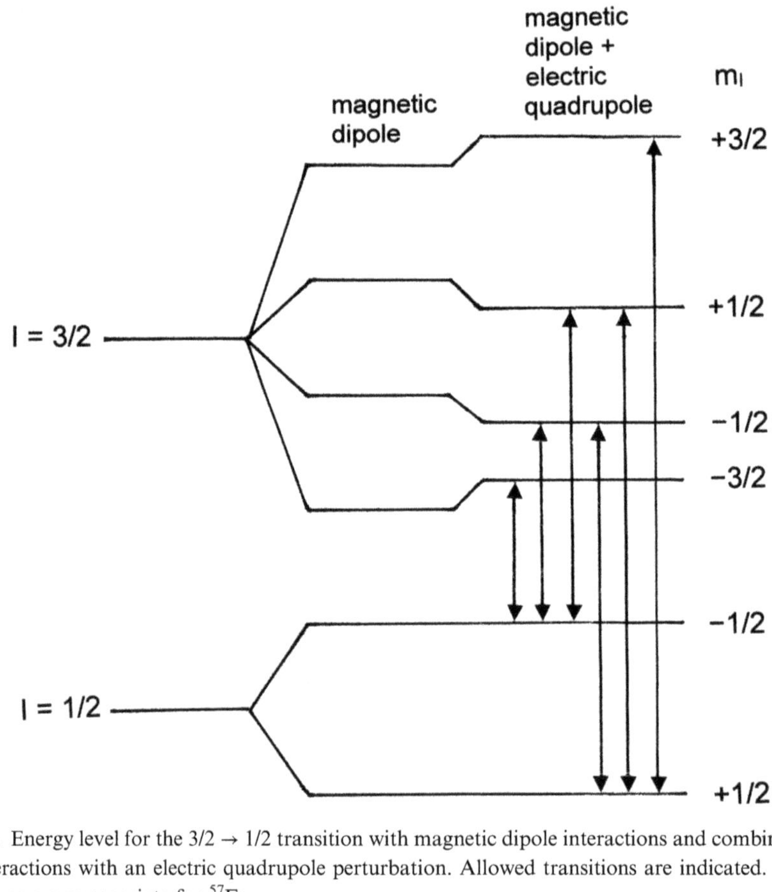

Figure 7.5. Energy level for the 3/2 → 1/2 transition with magnetic dipole interactions and combined magnetic dipole interactions with an electric quadrupole perturbation. Allowed transitions are indicated. The signs of the m_I values are appropriate for ^{57}Fe.

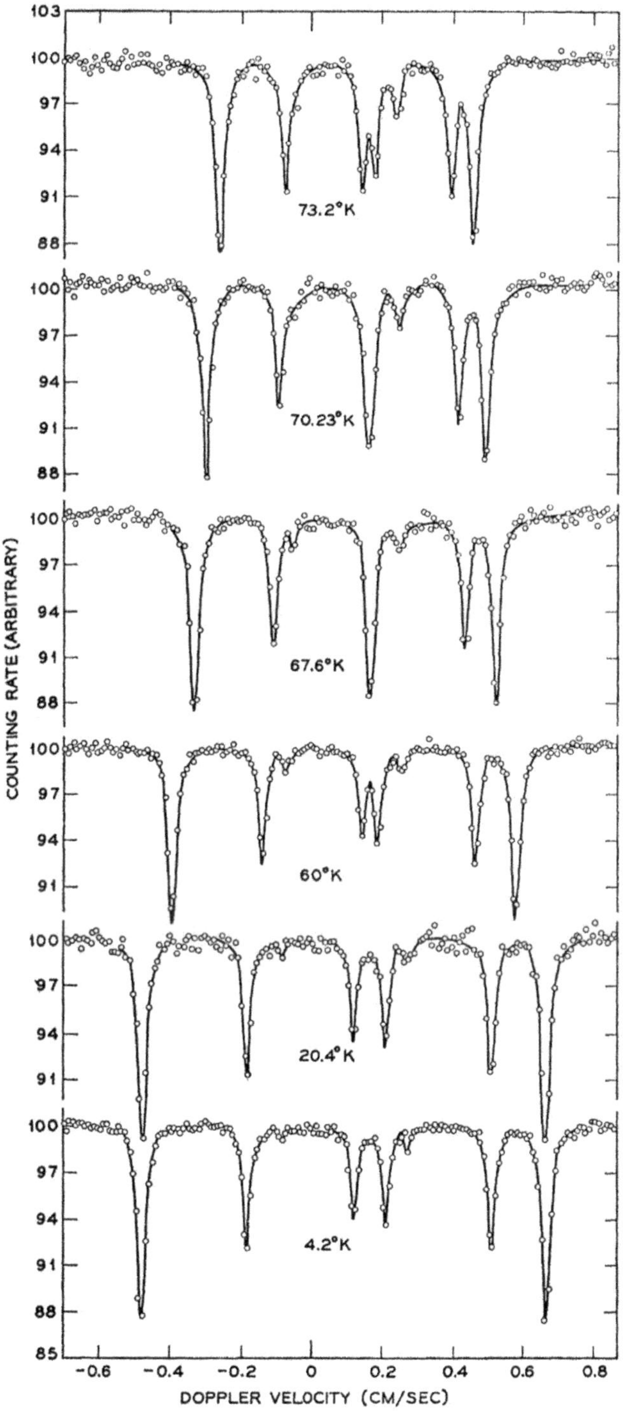

Figure 7.6. ^{57}Fe Mössbauer effect spectra of antiferromagnetic FeF_2 as a function of temperature below the Néel temperature. Reprinted with permission from Wertheim and Buchanan (1967). Copyright 1967 by the American Physical Society.

7.6 Problems

7.1. The Mössbauer transition for ^{121}Sb is a 5/2 →7/2 transition. Sketch (not necessarily to scale) the allowed transitions for:
(a) electric quadrupole splitting; and
(b) magnetic dipole splitting.

7.2. The ground state and excited state magnetic moments for the 14.4 keV transition of ^{57}Fe are +0.0902 μ_N and −0.1547 μ_N, respectively. In the absence of the chemical isomer shift, calculate the velocity (in mm s^{-1}) for the +1/2→−1/2 transition for a magnetic splitting due to a hyperfine field of 18.0 T.

7.3. The magnetic moments of the ground state and excited state of the Mössbauer transition in ^{119}Sn are −1.041 μ_N and +0.67 μ_N, respectively. Calculate the ratio of the splitting between the third and fourth peaks to the splitting of the first and second peaks in the spectrum shown in figure 7.3.

Bibliography and suggestions for further reading

Dunlap R A and Jones D F 1982 A Mössbauer effect study of the Sn impurity site hyperfine field in the Heusler alloys Co_2MnZ (Z = Al, Ga, Si, Ge, Sn) *Phys. Rev.* B **26** 6013–8

Greenwood N N and Gibb T C 1971 *Mössbauer Spectroscopy* (London: Chapman and Hall)

Kalvius M and Kienle P (ed) 2012 *The Rudolf Mössbauer Story: His Scientific Work and Its Impact on Science and History* (Heidelberg: Springer)

The High Fin Sperm Whale 2017 *Mössbauer spectrum of ^{57}Fe* (https://en.wikipedia.org/wiki/Mössbauer_spectroscopy#/media/File:M%C3%B6ssbauerSpectrum57Fe.svg)

Wertheim G 1964 *Mössbauer Effect: Principles and Applications* (New York: Academic)

Wertheim G K and Buchanan D N E 1967 Temperature dependence of the Fe^{57} hfs below the Néel temperature *Phys. Rev.* **161** 478–82

Yoshida Y and Langouche G (ed) 2013 *Mössbauer Spectroscopy: Tutorial Book* (Heidelberg: Springer)

IOP Publishing

The Mössbauer Effect (Second Edition)

Richard A Dunlap

Chapter 8

Experimental aspects of Mössbauer spectroscopy

8.1 Introduction

Early Mössbauer effect experiments utilized equipment along the lines of that shown in figure 2.3, where the Doppler shift was introduced by means of the source or absorber being attached to a rotating disk. The disk was rotated at a constant speed and data were accumulated for some period of time. These data represented one data point on a plot such as figure 2.4. The speed (and/or direction) of the disk was changed to measure another data point. While this was an effective means of demonstrating the principles of the Mössbauer effect, it is an inefficient method of accumulating useful data for applications.

Most Mössbauer effect spectrometers in use today utilize a similar design, as illustrated in figure 8.1. This system consists of a means of introducing the Doppler shift of the source γ-rays (the drive system), a means of detecting the γ-rays and a means of collecting the data. These three components of the Mössbauer spectrometer are discussed in the next three sections.

8.2 Mössbauer spectroscopy drive system

Current Mössbauer drive systems incorporate an electromagnetic transducer (sometimes referred to as a double loudspeaker design). A typical transducer or linear motor is shown in figure 8.2 and is used to introduce a linearly varying velocity, v, which gives a linearly varying energy, E, of the γ-rays according to the Doppler shift

$$E = E_0\frac{v}{c}$$

where E_0 is the γ-transition energy and c is the speed of light. The transducer consists of two spring mounted coils attached to an armature and suspended in a magnetic field produced by permanent magnets. One coil, the drive coil receives an oscillatory

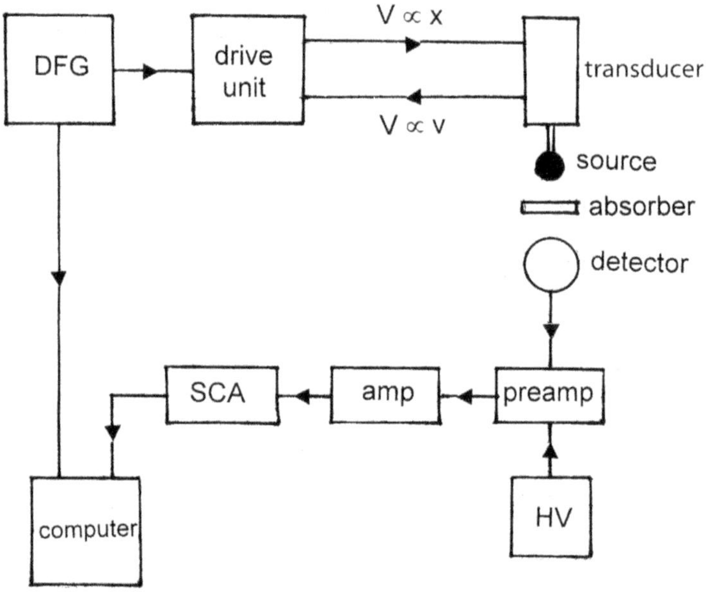

Figure 8.1. Block diagram of a Mössbauer spectrometer. DFG = digital function generator; HV = high voltage; SCA = single channel analyzer.

Figure 8.2. Typical electromagnetic Mössbauer transducer. The Mössbauer source is shown mounted on the right-hand end and is attached to an armature containing the drive coil and the feedback coil. The total length is about 14 cm. Image: Richard A Dunlap.

Figure 8.3. (a) Linear velocity reference signal and (b) parabolic drive signal.

voltage from the Mössbauer drive unit to drive the source back and forth through a linearly varying range of velocities. The periodicity of the drive is commensurate with the natural resonant frequency of the springs in the transducer and is typically in the range of about 20 Hz. If the spring behaves ideally according to Hooke's law, then the displacement of the coil and hence the source, x, will be proportional to the voltage, V, applied to it,

$$x \propto V. \tag{8.1}$$

The velocity of the source will be proportional to the time derivative of the position, or

$$v \propto \frac{dV}{dt}. \tag{8.2}$$

A linearly varying velocity represents a constant acceleration, a, or

$$a \propto \frac{d^2V}{dt^2}. \tag{8.3}$$

Thus, in this mode, the drive is often referred to as a constant acceleration drive. Since a in equation (8.3) is a constant and v in equation (8.2) varies linearly in time then x, and V, in equation (8.1) must have a parabolic time dependence. Each cycle of the drive signal will include a concave upward parabolic section followed by a concave downward parabolic section as illustrated in figure 8.3(b). This drive signal is produced by the drive unit in figure 8.1 using a reference signal from the digital function generator.

Since the transducer does not perfectly obey Hooke's law, it is necessary to adjust the shape of the drive signal to compensate for any non-linearities. This is the purpose of the second coil in the transducer. This coil generates a feedback voltage, V_{fb}, because of its motion in a magnetic field where

$$V_{fb} \propto v. \tag{8.4}$$

In order to observe non-linearities in the drive, the feedback voltage in equation (8.4) is compared to a perfect triangle wave (figure 8.3(a)) by the drive unit in figure 8.1

and the difference signal is used to adjust the parabolic drive signal (figure 8.3(b)) to ensure that the source velocity varies linearly.

8.3 Gamma-ray spectroscopy

As discussed in chapter 2, Mössbauer effect studies are undertaken on nuclei with relatively low transition energies. In fact, higher transition energies result in a reduction of the recoil-free fraction, and this reduces the ability to observe Mössbauer spectra. However, as noted in Greenwood and Gibb (1971) and Shenoy and Wagner (1978), there is interest in transitions up to around 150 keV. The energy of the transition is a major factor in the choice of γ-ray detector, although in some cases energy resolution may also be a consideration. For low energies (up to ~40 keV), gas-filled proportional counters are generally the choice. For intermediate energies (40–100 keV), gas-filled detectors have decreasing efficiency and a NaI/Tl scintillator crystal coupled to a photomultiplier tube is often the preferred choice. At higher energies a solid state detector is frequently used. In the early years of Mössbauer research, these were typically Li-drifted Ge detectors. These were inconvenient as the crystal had to be kept at liquid nitrogen, even when not in use. More recently intrinsic Ge detectors are commonly used, as these do not have to be kept at low temperature when the crystal is not biased. Ge detectors have good efficiency at higher energies, but also have the advantage of higher resolution and may find applications based on this property. For ^{57}Fe and ^{119}Sn Mössbauer studies, as have been described in some detail in the present book, gas-filled detectors offer reasonable efficiency and are cost effective and simple to use. The use of gas-filled proportional counters for Fe and Sn Mössbauer studies is considered in some detail below and this provides some insight into the approaches that can be taken to properly detect γ-rays for Mössbauer effect experiments.

Figure 8.4 shows a typical spectrum of a ^{57}Co source as obtained by a gas-filled proportional counter in the energy range up to about 17 keV. The location of the γ-rays from the ^{57}Fe Mössbauer transition is clearly seen. In addition, other peaks appear in the spectrum in this energy range. Figure 8.5 shows a simplified diagram of the processes during a Mössbauer event in an ^{57}Fe atom. The internal conversion peak is an x-ray peak that results from the Mössbauer decay energy liberating a K-shell atomic electron rather than emitting a γ-ray. The K shell binding energy is 7.1 keV and the electron that is liberated has the remaining $14.4 - 7.1 = 7.3$ keV of energy. This is followed by the emission of a characteristic x-ray of energy 6.4 keV when an L shell electron falls into the K shell vacancy. The Mössbauer escape peak results from Mössbauer γ-rays that have lost energy from liberating a K-shell electron from the detector medium, i.e., in this case, a Xe atom, and is not directly part of the process shown in figure 8.5. The escape peak, which is caused by the Mössbauer γ-rays, carries all the relevant information concerning the Mössbauer transition. In many cases it can be counted along with the Mössbauer peak in order to improve the count rate of the spectrometer.

The signal from the proportional counter is processed by a single channel analyzer in order to choose pulses corresponding to the proper energy and reject

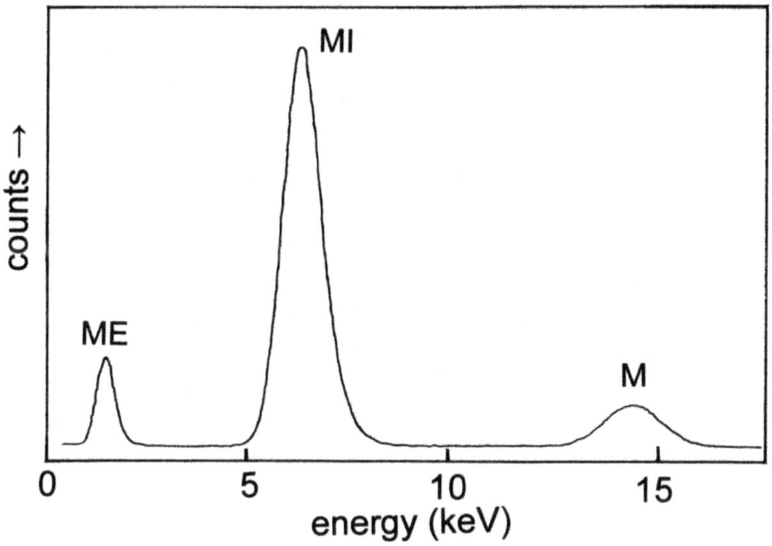

Figure 8.4. Energy spectrum of a ^{57}Co radioactive source obtained with a Xe:CO_2 proportional counter. M = Mössbauer peak, MI = Mössbauer internal conversion (x-ray) peak and ME = Mössbauer escape peak.

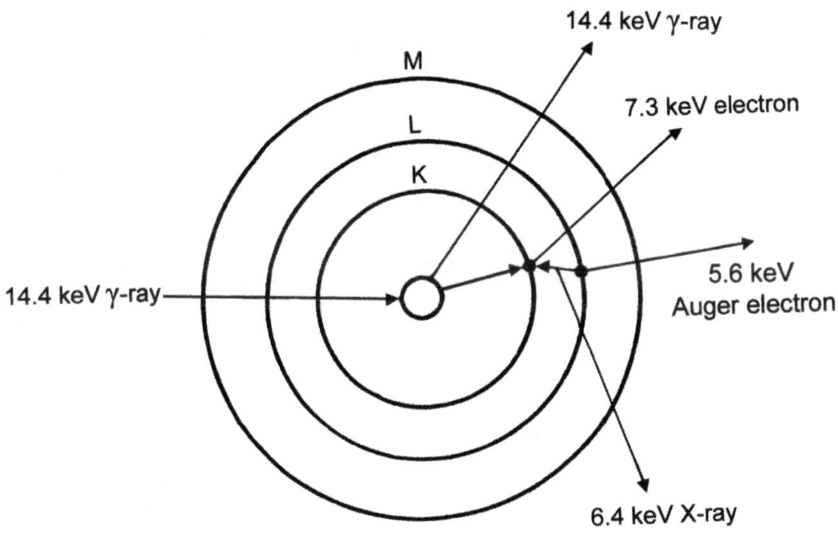

Figure 8.5. Simplified diagram of the absorption and re-emission of a 14.4 keV γ-ray from an ^{57}Fe nucleus and the associated processes. Auger electrons are not discussed in detail.

other pulses. Some analyzers, particularly those intended for use for Mössbauer effect spectrometry, can discriminate on two energies corresponding to the Mössbauer effect peak and the Mössbauer escape peak.

The spectrum of a ^{119m}Sn source as obtained by a gas-filled proportional counter is shown in figure 8.6(a). In this case the Mössbauer peak and an internal conversion

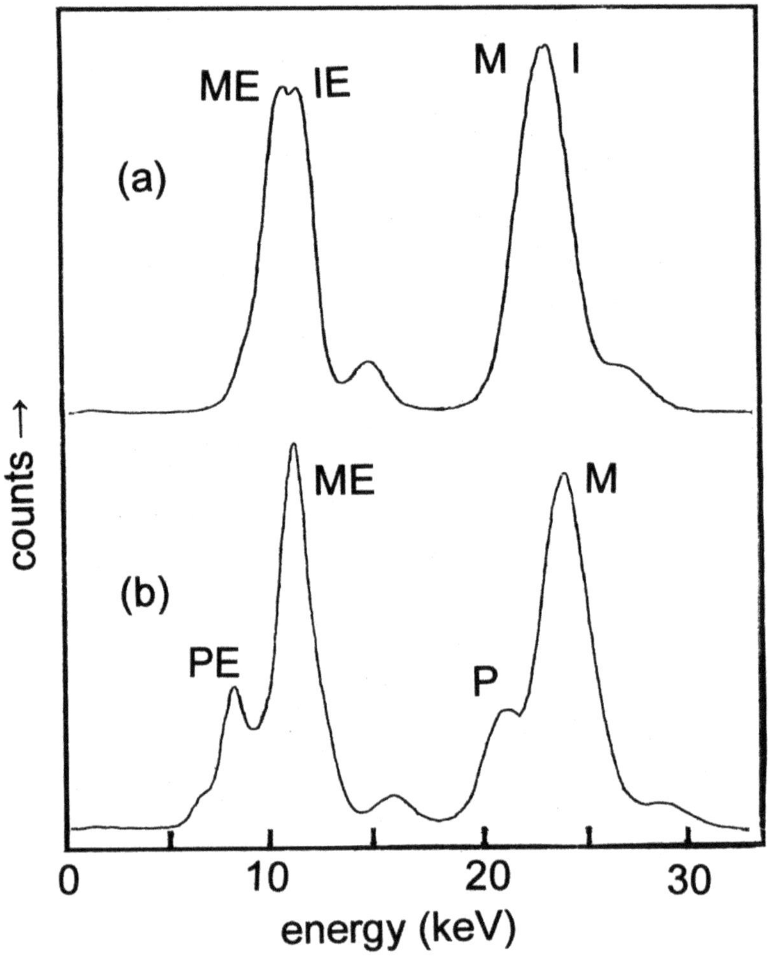

Figure 8.6. (a) Spectrum of a ^{119m}Sn source obtained using a Kr:CO_2 proportional counter and (b) spectrum of a ^{119m}Sn source obtained using a Kr:CO_2 proportional counter with a Pd filter. In (a) I = internal conversion peak, M = Mössbauer peak, IE = internal conversion escape peak, and ME = Mössbauer escape peak. In (b) M = Mössbauer peak, P = palladium x-ray, ME = Mössbauer escape peak and PE = palladium x-ray escape peak.

peak cannot be resolved by the proportional counter. The internal conversion peak in this spectrum can be understood by an inspection of the ^{119m}Sn decay scheme shown in figure 2.10. Internal conversion of the γ-rays from the decay of the 89 keV ^{119m}Sn state to the 23.8 keV Mössbauer state, produces Sn K-shell x-rays at an energy of 29.2 keV. Although the proportional counter resolution is not sufficient to distinguish these two peaks, it is straightforward to filter out the higher energy internal conversion peak, as shown in figure 8.6(b). Palladium has a K-absorption edge at 24.35 keV and a Pd foil will absorb the 29.2 keV internal conversion x-rays but not the 23.8 keV Mössbauer γ-rays. This also eliminates the lower energy internal conversion escape peak, although it introduces some small additional palladium x-ray peaks. It is, however, possible to discriminate the relevant Mössbauer peaks and eliminate the x-rays from unwanted processes.

8.4 Data accumulation

Traditionally Mössbauer data were collected using a dedicated multichannel analyzer. In recent years, the Mössbauer transducer has been coupled to a data collection computer, as illustrated in figure 8.1. A digital function generator (DFG) provides a signal to the Mössbauer drive unit which interfaces with the transducer. The function generator also sends a signal to a computer to convey information about the transducer velocity as a function of time. The computer counts and stores information from the γ-ray detector about the γ-ray flux. Data are stored during constant duration time windows throughout the velocity scan, each window corresponding to a specific velocity window. As the velocity is swept through its specified range, the computer accumulates data for the appropriate velocity window. The number of data points is typically a power of 2, e.g., 512, 1024, etc. Typical Mössbauer data consists of two mirror image spectra corresponding to the increasing and decreasing components of the linearly varying velocity curve shown in figure 8.3. Data collection can utilize a plug-in board which allows the computer to function as a multichannel analyzer. As well, several manufacturers produce standalone units, primarily designed for Mössbauer spectroscopy, that serve as multichannel analyzers and connect to a computer via a USB port.

8.5 Velocity calibration

Since each data point of a Mössbauer spectrum as collected corresponds to a certain range of Doppler velocities, it is essential to be able to determine the relationship between channel number in the spectrum and the actual velocity. This will also allow for the two mirror image spectra as collected to be folded onto each other. There are several approaches that have been taken to calibrating Mössbauer effect spectra. Most commonly an additional spectrum is obtained using the same Mössbauer source and transducer settings with an absorber that has spectral features at known velocities. An absorber with at least two peaks in its spectrum is needed for such purposes. For ^{57}Fe Mössbauer effect spectroscopy, room temperature α-Fe is universally used as a calibration standard. α-Fe is ferromagnetic at room temperature with a hyperfine field of 33.0 T and has a spectrum that is a Zeeman split sextet, as illustrated in figure 7.2. The known velocities of the six peaks in the room temperature α-Fe spectrum are given in table 8.1. Assigning these velocities to the fitted locations of the peaks in the Mössbauer spectrum allows for the determination of the velocity associated with each channel of the experimental spectrum.

The choice of a calibration standard for ^{119}Sn Mössbauer spectroscopy is not so straightforward. This is because pure tin (i.e., the stable room temperature phase, β-Sn) has a ^{119}Sn Mössbauer effect spectrum that is a single line. The use of a tin-containing material with more than one peak in the Mössbauer spectrum can be used as a calibration standard. An example is shown by the magnetically split Sn spectrum in figure 7.3. However, a sandwich of pure Sn with $BaSnO_3$ is also

Table 8.1. Velocities of the six hyperfine split Mössbauer lines of α-Fe at room temperature.

Line	Velocity (mm s^{-1})
Fe 1	−5.3123
Fe 2	−3.0760
Fe 3	−0.8397
Fe 4	+0.8397
Fe 5	+3.0760
Fe 6	+5.3123

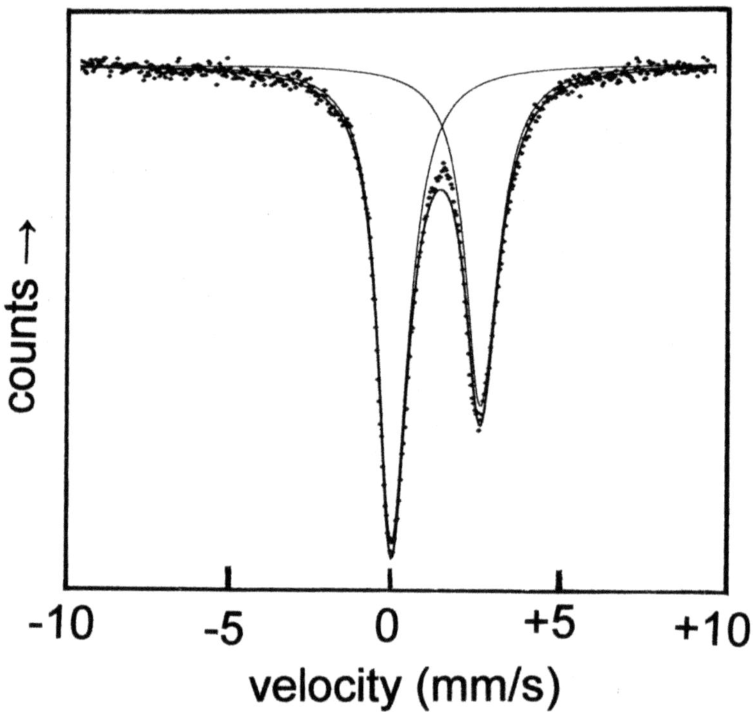

Figure 8.7. Room temperature ^{119}Sn Mössbauer effect spectrum of an absorber of β-Sn/$BaSnO_3$. The thickness of each absorber was adjusted to (more or less) compensate for the differences in recoil-free fraction of the different materials at room temperature.

commonly used. $BaSnO_3$ has a Mössbauer spectrum that is a singlet, but the isomer shift is substantially different than for pure Sn and this gives a Mössbauer spectrum with two peaks, as shown in figure 8.7. The known velocities of the ^{119}Sn Mössbauer peaks for β-Sn and $BaSnO_3$ are given in table 8.2, and these values allow for the calibration of a ^{119}Sn Mössbauer spectrum.

Table 8.2. Mössbauer parameters for an absorber of Sn/$BaSnO_3$.

Line	Velocity (mm s^{-1})
$BaSnO_3$	0.000
Sn	2.540

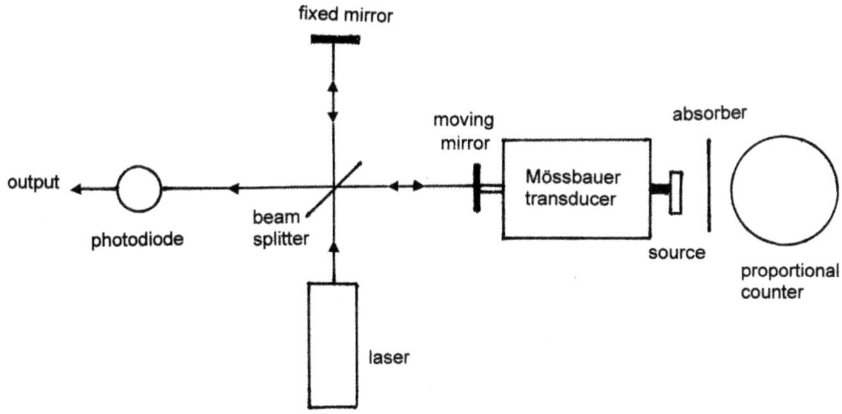

Figure 8.8. Basic design of a laser interferometer for Mössbauer drive calibration.

Another approach to calibrating Mössbauer spectra is the use of an additional ^{57}Fe Mössbauer source mounted on the back end of the Mössbauer velocity transducer and the simultaneous collection of an α-Fe Mössbauer calibration spectrum. This approach may be useful for Mössbauer experiments involving nuclides for which there is no convenient calibration standard. This is also sometimes a convenient option for backscatter experiments (see section 8.8) where a conventional transmission geometry experiment can provide a calibration spectrum simultaneously. This approach has been used for the Mössbauer spectrometer that was utilized on the Martian surface (see section 10.4).

A final method of Mössbauer spectrum calibration is the use of a laser interferometer attached to the back end of the velocity transducer. A basic diagram of this arrangement is illustrated in figure 8.8. As the source vibrates back and forth, so does the moving mirror. This introduces periodic interference fringes in the output of the interferometer which are observed by the photodiode. The output of the photodiode can then be related to the source velocity. This approach has the advantage that the drive calibration (and hence stability) can be monitored continuously throughout data accumulation. It is also applicable to all velocity ranges and may be a suitable choice for Mössbauer nuclides that require large velocities. On the downside, it may be an expensive option for small Mössbauer facilities or the use of Mössbauer effect spectroscopy as a teaching tool in an undergraduate laboratory.

Table 8.3. Common isomer shift references for Mössbauer transitions. Typically used at room temperature.

Mössbauer nuclide	Isomer shift reference
^{57}Fe	α-Fe
^{119}Sn	$BaSnO_3$

An important aspect of Mössbauer spectrum calibration is the establishment of a standard for isomer shift measurement. In the spectrum as accumulated, the isomer shift is determined for the sample relative to the source. As different experiments frequently use sources with different host matrices, the source details need to be specified in order to provide meaning for the measured isomer shifts. It is, therefore, customary to reference measured isomer shifts for a given Mössbauer nuclide to the isomer shift of a standard reference material. The commonly accepted isomer shift references for ^{57}Fe and ^{119}Sn are given in table 8.3. Some degree of standardization of isomer shift references has been established for most other Mössbauer effect nuclides and Stevens and Gettys (1978) provide relevant information for many nuclides.

8.6 Data analysis

Once Mössbauer spectral data have been collected they can be calibrated and folded to align the two mirror image spectra as described above. This process is typically accomplished by the Mössbauer fitting software. A number of commercial, as well as freeware, Mössbauer fitting programs are available for fitting spectra to appropriate models. While the appropriate model for the analysis of a Mössbauer spectrum depends on the specific material being investigated, most fitting functions are based on the Lorentzian, as discussed in chapter 2. The basic functional form of the Lorentzian line as a function of Doppler velocity, v, is

$$L(v) = \frac{A}{\Gamma^2 + (v - v_0)^2}$$

where A/Γ^2 is the maximum amplitude of the resonance, v_0 is the center of the resonance curve and Γ is the half width at half maximum of the resonance curve in velocity units. For an absorption spectrum the intensity is measured below the background level. As described in previous chapters, Mössbauer spectra consist of several resonance peaks as appropriate for the details of the specific Mössbauer transitions and the hyperfine interactions characteristic of the absorber material. In the simple case, the total spectrum is the sum of individual Lorentzian peaks, as appropriate, and is expressed as

$$L(v) = \sum_i^N L_i(v) = \sum_i^N \frac{A_i}{\Gamma_i^2 + (v - v_{0i})^2}. \tag{8.5}$$

Appropriate constraints on the widths and amplitudes, A_i, of the various peaks are included in the fitting procedure. Typically, in the number of Lorentzians included in the fit, N, for ^{57}Fe or ^{119}Sn would be two for an electric quadrupole transition and six for a magnetic dipole transition. Spectra may also have multiple patterns corresponding to multiple crystallographic sites in the structure or multiple crystallographic phases in the sample.

In many cases, chemical or structural disorder causes broadening of the spectrum, and it is necessary to introduce a suitable functional form of the spectral lines to account for this. In general, such spectra are fit to a function based on the convolution of the sum of the intrinsic Lorentzian lines, L, with a distribution, P, of appropriate hyperfine parameters. This is expressed as

$$I(v) = \int P(v')L(v - v')dv'$$

where $I(v)$ is the total intensity relative to the background as a function of Doppler velocity. The Lorentzian is expressed in terms of the parameters given in equation (8.5). This approach can be used for quadrupole split spectra, where the quadrupole distribution, $P(\Delta)$, follows from the velocity distribution. For Zeeman split spectra, the hyperfine field distribution, $P(H)$, is related to the velocity distribution in accordance with the field–velocity relations given in section 7.3. In the discussion below, the generic distribution, $P(v)$ is used. Specific examples related to quadrupole splitting and hyperfine field distributions are given (for example) in sections 9.5 and 10.3.

An important aspect of this type of analysis is the ability to extract the probability distribution, $P(v)$ from the spectrum. This distribution is a measure of the distribution of Mössbauer probe atom environments in the sample. In the case of quardupole split spectra, this probability distribution reflects the variations in local symmetry around the probe atoms and in the case of Zeeman split spectra, it reflects the variations in the magnetic environments.

Various approaches to the modelling of the probability distribution have been utilized. The basic approach is to use a functional form for $P(v)$ that can approximate the shape of the actual distribution function, and which has parameters to describe this shape that can be readily fit to the data. Some of these include the use of a series of discrete values, Fourier expansions and Voigt-based functions. These are described briefly below.

Discrete functions. The probability distribution may be expressed as a series of N step functions, $W_i(v')$, (e.g., Hesse and Rübartsch 1974, Le Caër and Dubois 1979) as

$$P(v') = \sum_{i}^{N} a_i W_i(v')$$

where the width of each step is $v_{\max}/N$. The N amplitudes of the step functions, a_i, are fitted parameters that give the functional form of $P(v)$.

Fourier expansions. Window (1971) has suggested the use of a Fourier expansion to describe the probability distribution. This is expressed as

$$P(v') = \sum_{i}^{N} a_i[\cos(i\pi v'/v_{\max}) - \cos i\pi]$$

where the Fourier amplitudes, a_i, are fitted parameters.

Voigt-based functions. A Voigt function is the convolution of a Lorentzian and a Gaussian. Thus, in equation (8.5), the probability distribution is represented by a sum of Gaussians of the form (e.g., Rancourt and Ping 1991)

$$P(v') = \sum_{i}^{N} \frac{a_i}{\sqrt{2\pi\sigma_i^2}} \exp\left[-\frac{(v' - v_i)^2}{2\sigma_i^2}\right]$$

where each Gaussian component is described by a fitted width, σ_i, amplitude, a_i, and peak velocity, v_i.

In all cases as described above, the choice of the number of components is important for the proper description of the probability distribution. Using too few components in the fit can fail to reproduce relevant features in the distribution, while the use of too many components can result in the appearance of unphysical artifacts.

Disordered microstructures that give rise to distributions of quadrupole splittings or hyperfine magnetic fields typically also give rise to a distribution of isomer shifts. One way of dealing with the isomer shift distribution is to assume that it is correlated with the hyperfine distribution. This correlation is expressed as

$$\delta = \delta_0 + \alpha\Delta$$

or

$$\delta = \delta_0 + \alpha H$$

for quadrupole splitting and hyperfine field distributions, respectively. In these expressions, δ_0 and α are fitted parameters. Alternately, as illustrated by the example in section 10.3, the distribution of quadrupole splittings or hyperfine fields and the distribution of isomer shifts can be treated independently.

8.7 Temperature control

While the majority of Mössbauer measurements reported in the literature have been made at room temperature, there are reasons why measurements at other temperatures can be of interest. Firstly, as noted above, Mössbauer nuclides with transition energies that are higher than those of (e.g.,) ^{57}Fe and ^{119}Sn, can benefit from increasing the recoil-free fraction by lowering the temperature of either the absorber or source or both. Secondly, the absorber may exhibit interesting phenomena at temperatures other than room temperature. In this latter case, temperatures either below room temperature or above room temperature or both may be of interest. In the brief overview given in this section, we consider some appropriate techniques for conducting Mössbauer effect studies at low and elevated temperatures.

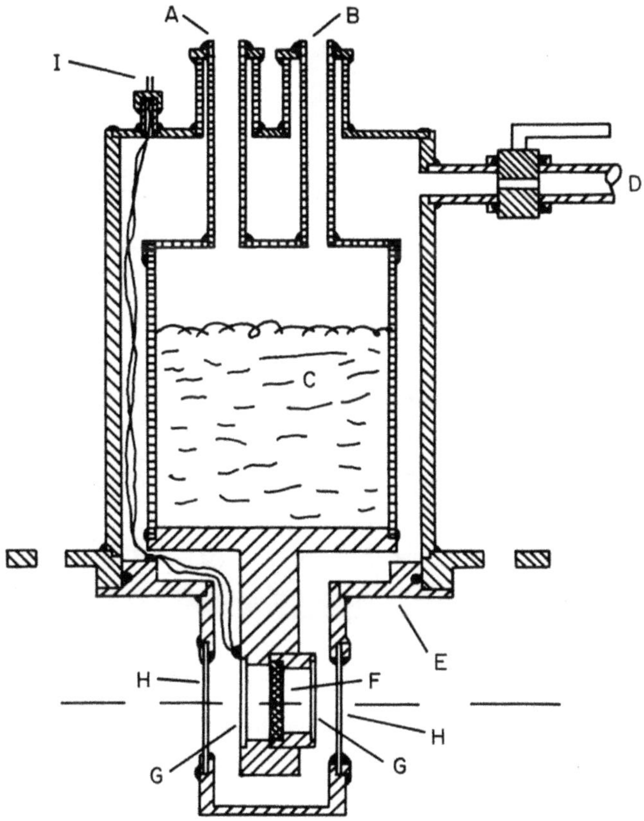

Figure 8.9. Diagram of a simple nitrogen cryostat for cooling a Mössbauer absorber using liquid nitrogen, (A) liquid nitrogen fill port, (B) nitrogen vent, (C) liquid nitrogen, (D) vacuum port, (E) detachable tail section, (F) Mössbauer absorber, (G) aluminum foil, (H) mylar window and (I) thermocouple feed through. Reprinted from Benczer-Koller and Herber (1968). Copyright (1968). With permission from Elsevier.

Figure 8.9 shows the design of a simple cryogenic system for lowering the temperature of a Mössbauer absorber to near liquid nitrogen temperature, while maintaining the source at room temperature. This arrangement can improve the recoil-free fraction of the absorber and also allows for the study of the properties of the absorber at low temperature. Some degree of temperature control can be achieved by incorporating a heater coupled to the absorber holder. The velocity transducer is positioned to allow the γ-rays to travel horizontally through the mylar windows to the detector of the other side.

Figure 8.10 shows a helium cryostat designed for Mössbauer effect measurements. The main diagram illustrates a system with the capabilities of cooling both the absorber and the source to liquid helium temperature. This can substantially improve the total recoil-free fraction and allows for the study of Mössbauer nuclides with large transition energies which cannot be investigated at room temperature. The source is attached to a long drive rod and is connected to the velocity transducer

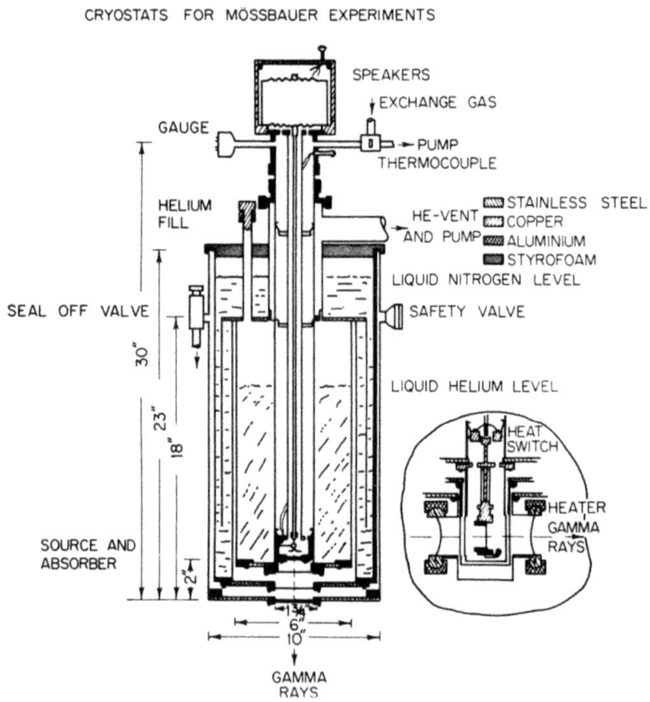

Figure 8.10. Diagram of a helium cryostat for absorber or absorber/source cooling. Reprinted from Benczer-Koller and Herber (1968). Copyright (1968). With permission from Elsevier.

which is at room temperature above the cryostat. The γ-rays travel vertically through the absorber and then the windows (mylar of beryllium) at the bottom of the cryostat to the detector is positioned below the cryostat.

The small diagram shows the design of a tail section where the geometry of the velocity transducer and detector is the same as for the nitrogen cryostat in figure 8.9. In this case, the absorber is cooled while the source is held at room temperature.

Measurements in an applied magnetic field can be accomplished using a system similar to that in figure 8.10, by the incorporation of a superconducting solenoid in the region around the absorber. When the source is cooled using a vertical drive rod, it may be necessary to incorporate cancellation coils to ensure that the magnetic field at the source is zero. For the horizontal drive design with a room temperature source, a split superconducting solenoid pair can be utilized to allow for a γ-ray beam path through the cryostat.

In recent years, the use of closed-cycle helium refrigerators for low temperature Mössbauer measurements has become common. A typical design is illustrated in figure 8.11. Cooling is accomplished by means of the Gifford–McMahon refrigeration cycle (Gifford 1966). Helium gas is compressed by a compressor and cooling takes place in the expander shown in the lower portion of the figure. The expander is connected to the compressor by two gas lines, one to carry high pressure helium gas from the compressor and one to return low pressure helium gas to the compressor.

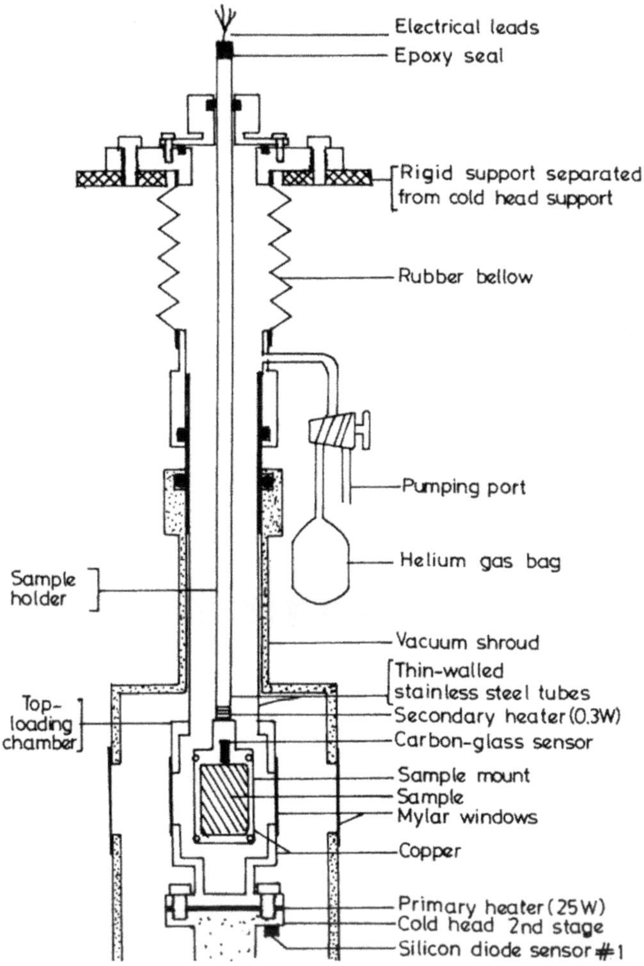

Figure 8.11. Diagram of a closed-cycle helium refrigerator for Mössbauer applications. From Senthil Kumar *et al* (1996), copyright IOP Publishing. Reproduced with permission. All rights reserved.

These systems are convenient, as they are charged with helium gas and do not require the transfer of liquid helium each time they are used. Temperature control is straightforward and sample temperature is adjustable from around liquid helium temperature (or lower) to room temperature (or higher). Typically, samples can be changed easily and quickly (compared to liquid helium cryostat systems). Mössbauer measurements utilize an external room temperature source and velocity transducer with access through mylar windows, as shown in the figure.

A major consideration for the use of a closed-cycle refrigerator for Mössbauer effect measurements is the vibration produced by the helium compressor. These vibrations contribute to the linewidth of the Mössbauer transition and degrade the quality of the resulting spectra. It is, therefore, essential to suitably isolate the Mössbauer system, including the absorber, from any vibrations produced by the

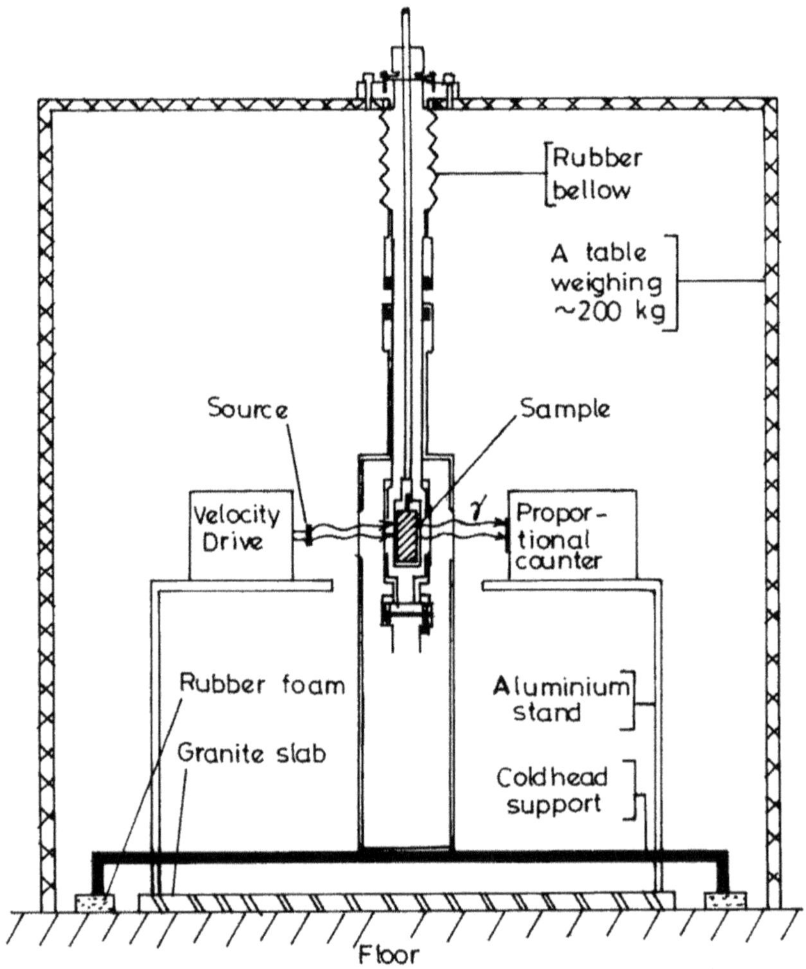

Figure 8.12. System for isolating the Mössbauer drive and absorber from laboratory vibrations caused by the helium compressor in a closed-cycle helium refrigerator used for Mössbauer effect spectroscopy. From Senthil Kumar *et al* (1996), copyright IOP Publishing. Reproduced with permission. All rights reserved.

refrigeration system. As illustrated in figure 8.11, the cold portion of the refrigerator is isolated from the warm portion of the refrigerator by a rubber bellows. Figure 8.12 shows a typical approach to isolating the Mössbauer system from vibrations produced by the compressor. The top portion of the refrigerator (above the rubber bellows) is anchored to the laboratory floor by a heavy streel frame and the lower portion of the refrigerator (below the rubber bellows) and the Mössbauer drive system is isolated from the floor by rubber foam mounts.

Various designs of Mössbauer ovens have appeared in the literature. A typical example is shown in figure 8.13. The design consists of an evacuated tube where the Mössbauer absorber can be heated. Appropriate temperature measurement and control can be incorporated into the system. The Mössbauer velocity transducer and

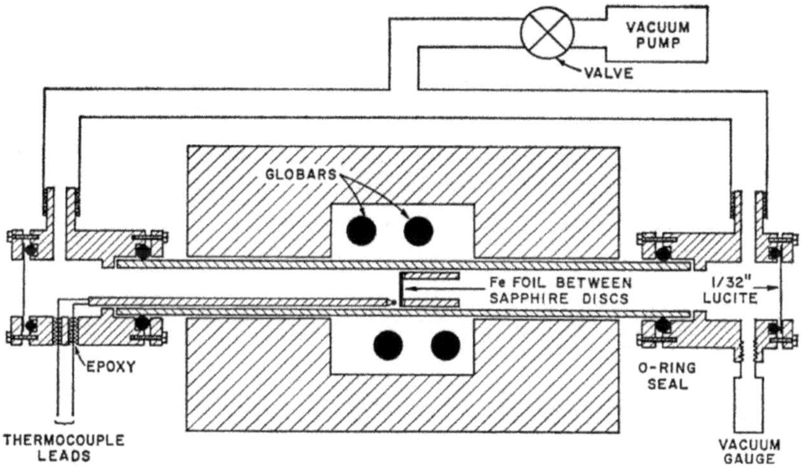

Figure 8.13. Design of a Mössbauer oven. Reprinted with permission from Kovats and Walker (1969). Copyright 1969 by the American Physical Society.

source, as well as the γ-detector, are outside of the oven and are at room temperature. The γ-rays enter and exit the oven through thin plastic windows.

8.8 Transmission and backscatter geometries

The most commonly utilized geometry for Mössbauer effect spectroscopy is the transmission geometry, as illustrated in figure 8.1. In this geometry the absorber to be studied is placed between the source, which is attached to the velocity transducer, and the detector. When the condition for recoil-free nuclear resonance fluorescence is met by the Doppler shifted γ-rays from the source, then there will be a net decrease in the radiation reaching the detector. This is because, as described in chapter 1, the resonantly re-emitted γ-rays are distributed isotropically. As the velocity of the source is scanned through the resonance condition, the characteristic Lorentzian line shape, as given by equation (2.5), is observed. The preparation of the absorber is an important aspect of obtaining a suitable Mössbauer spectrum. Particularly, the thickness of the absorber must be considered. As the thickness of the absorber is increased, the number of Mössbauer active nuclei will increase and the amount of observed absorption in the spectrum will increase, at least up to some point. At some thickness, resonant γ-rays emitted by one Mössbauer nucleus will be significantly absorbed by other Mössbauer nuclei before leaving the absorber. This process introduces broadening of the Mössbauer spectral peaks. For a greater thickness, the actual number of resonant γ-rays leaving the absorber will decrease. The so-called ideal thickness of the absorber will provide the maximum resonant absorption without these undesirable effects. Table 8.4 gives the ideal absorber thicknesses in terms of the density of the Mössbauer element per unit area. As an example, we can consider the ^{57}Fe Mössbauer measurement of a 2 cm diameter absorber of Fe_2O_3. Using the thickness from the table, we obtain a total absorber mass of 72 mg.

Table 8.4. Ideal absorber thickness for transmission geometry Mössbauer effect spectroscopy for different Mössbauer nuclides. Thickness is given as mass of naturally occurring element per unit area.

Mössbauer nuclide	Ideal thickness
^{57}Fe	16 mg Fe/cm^2
^{119}Sn	13 mg Sn/cm^2
^{121}Sb	40 mg Sb/cm^2
^{125}Te	32 mg Te/cm^2

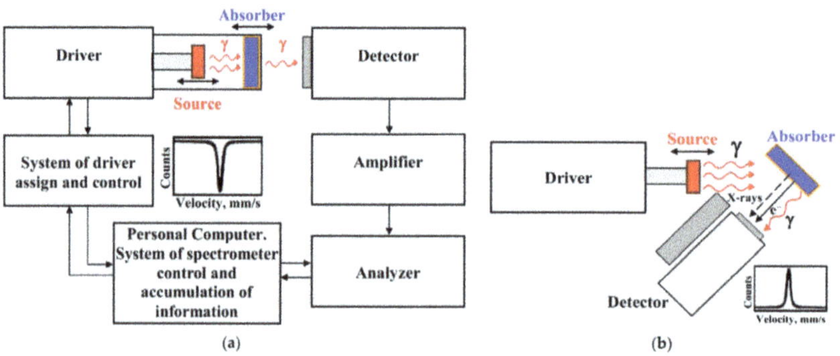

Figure 8.14. Comparison of (a) transmission and (b) backscatter geometries for the collection of Mössbauer effect spectra. For the backscatter geometry the data collection components of the system are not shown. From Maksimova and Oshtrakh (2021) Copyright by the authors. CC BY 4.0

An alternative to the transmission geometry, as described above, is the backscatter geometry. A comparison of these two approaches is shown in figure 8.14. In the backscatter geometry, resonant γ-rays are re-emitted isotropically and those that travel back towards the source are detected, as shown in the figure. Several detectors can be placed annularly around the source in order to improve the count rate from backscattered radiation. In this geometry, the γ-rays that are detected are only those that are resonantly backscattered. Thus, the count rate will increase when the Doppler velocity of the source goes through resonance, as illustrated by the insert in figure 8.14.

It is possible to observe the resonance by detecting either Mössbauer γ-rays, Mössbauer internal conversion x-rays or Mössbauer internal conversion electrons. In the case of the 14.4 keV Mössbauer transition in ^{57}Fe, figure 8.5 provides information about these radiations. Traditional backscatter Mössbauer effect spectroscopy utilizes the 14.4 keV γ-rays or 6.4 keV x-rays and the illustration in figure 8.14 shows the detection of these two radiations. The use of the internal conversion electrons for Mössbauer effect spectroscopy is discussed in detail in the next section.

In view of the constraints placed on the form of absorbers for transmission Mössbauer effect experiments, the use of backscatter measurements can be an attractive alternative when preparing suitable absorbers is undesirable (for example if non-destructive analysis is necessary) or even impossible. Some possible situations that can fall into this category are studies of mineral specimens and the study of structural materials. The *in situ* study of Martian minerals as described in section 10.4 is an example of the former situation. The study of bulk terrestrial mineral samples is also sometimes conveniently carried out using the backscatter geometry. As an example of the latter category of measurements, figure 8.15 shows the results of an ^{57}Fe Mössbauer investigation of a weathered steel plate. Figure 8.15(a) shows

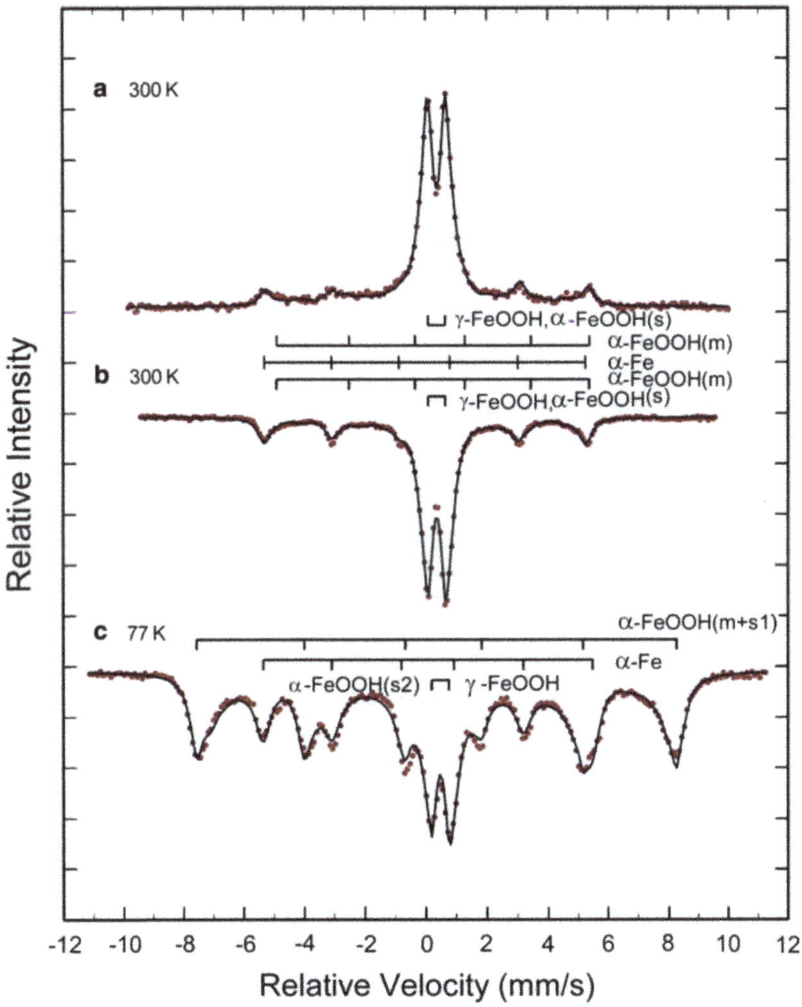

Figure 8.15. ^{57}Fe Mössbauer effect spectra of weathered steel samples showing the presence of various iron oxides. (a) X-ray backscatter spectrum at 300 K, (b) transmission spectrum at 300 K and (c) transmission spectrum at 77 K. Reprinted from Cook (2005). Copyright (2005). With permission from Elsevier.

a backscatter measurement of the oxidized surface of the steel plate, along with the identification of the various iron oxides that are present in the surface layer. As the escape depth of the photons is limited, this technique preferentially samples the surface region of the bulk sample. In the case of the spectrum obtained with 6.4 keV internal conversion x-rays, the escape depth is about 10 μm. For comparison, figure 8.15(b) shows a transmission geometry Mössbauer effect spectrum of an absorber prepared from oxide material scraped from the surface of the weathered steel sample. A low temperature transmission spectrum is shown in figure 8.15(c).

8.9 Internal conversion electron Mössbauer spectroscopy

Referring to figure 8.5, we see that γ-decay and internal conversion are competing processes for the de-excitation of the 14.4 keV state of ^{57}Fe. The same is true of the 23.8 keV decay of ^{119}Sn, as shown in figure 8.16. In general, the internal conversion coefficient is defined as the ratio of decays that proceed by emission of an internal conversion electron to those that proceed by the emission of a γ-ray (Dunlap 2023). Table 8.5 gives values of the internal conversion coefficients for the 14.4 keV decay of ^{57}Fe and the 23.8 keV decay of ^{119}Sn. The energy of the internal conversion electrons from the 14.4 keV decay of ^{57}Fe and the 23.8 keV decay of ^{119}Sn are given in table 8.6. It is clear that, in both cases, the decay process produces more internal conversion electrons than γ-rays. This feature suggests that the study of internal conversion electrons from Mössbauer events can provide useful information. This approach is referred to as conversion electron Mössbauer spectroscopy (CEMS).

It is important to consider the differences between electrons and γ-rays in order to understand how to utilize conversion electrons in a Mössbauer experiment. Most importantly, the range of electrons in matter is substantially less than the range of photons of similar energy. For example, the range of internal conversion x-rays in a

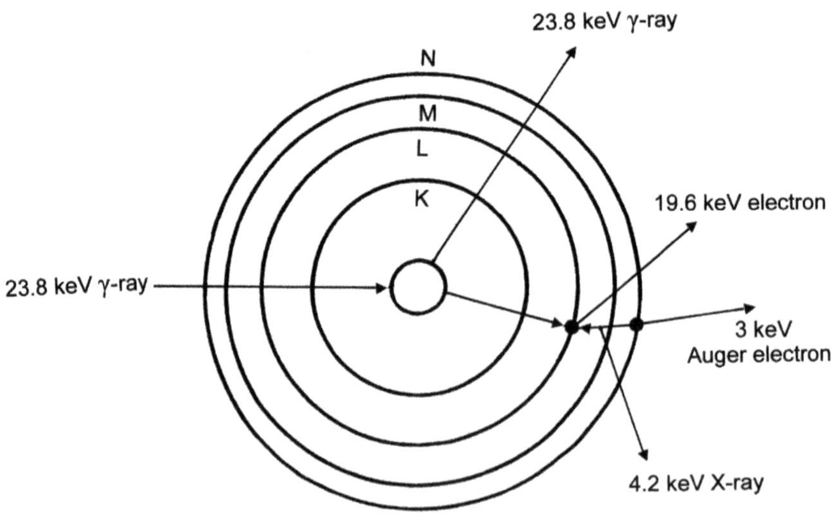

Figure 8.16. Simplified diagram of the absorption and re-emission of a 23.8 keV γ-ray from a ^{119}Sn nucleus and the associated processes. Auger electrons are not discussed in detail.

Table 8.5. Internal conversion coefficients, α, for the Mössbauer transitions of ^{57}Fe and ^{119}Sn.

Nuclide	γ-Ray energy (keV)	α	References
^{57}Fe	14.4	8.17	Rubinson and Gopinathan (1968)
^{119}Sn	23.8	5.13	Kostroun and Crasemann (1968)

Table 8.6. Energy of internal conversion electrons for the Mössbauer transitions of ^{57}Fe and ^{119}Sn.

Nuclide	Electron energy (keV)
^{57}Fe	7.3
^{119}Sn	19.6

typical Mössbauer absorber might be tens of μm while the range of internal conversion electrons might be a hundred nm. For this reason, the Mössbauer sample is often contained within the electron detector so that the electrons are not attenuated by the detector window.

A variety of different detector designs have been incorporated into conversion electron Mössbauer experiments. A typical example is illustrated in figure 8.17. Gamma-rays from the source reach the sample through a mylar window. A proportional counter with continuous gas flow is used as the detector. The sample under investigation serves as the cathode and two thin wires serve as the anode.

An important feature of conversion electrons is that they lose energy as they travel from the point of emission to the sample surface. This means that the energy of an electron is a measure of the depth below the surface where the electron was emitted. This feature means that conversion electron Mössbauer spectroscopy is depth selective. Measuring the Mössbauer spectrum for electrons of a specific energy provides information about the properties a specific distance below the surface. A Mössbauer spectrometer designed to discriminate electrons on the basis of their energy is illustrated in figure 8.18. In this design, the chamber is evacuated to eliminate the loss of energy as the electrons travel from the sample to the detector. The detector is a scintillation crystal coupled to a photomultiplier tube through a light pipe. As the detector is only sensitive to electrons with energies greater than about 10 keV, the system is appropriate for ^{119}Sn Mössbauer spectroscopy (see table 8.6). The electron energy that is measured, and hence the depth in the sample, is controlled by the magnetic field produced by the coils. The application of this technique to the study of depth dependent phenomena is discussed in section 9.7.

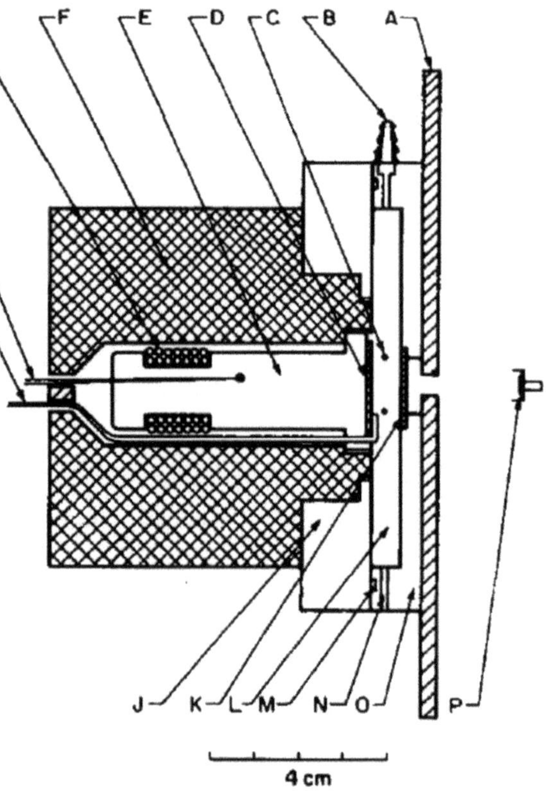

Figure 8.17. Detector assembly for conversion electron Mössbauer effect spectroscopy incorporating a heater for high temperature measurements, (A) lead shielding and γ-ray collimator, (B) gas inlet, (C) anode, (D) sample (as cathode), (E) copper rod (as hot finger), (F) Teflon thermal insulator, (G) heat resistor, (H) control thermocouple, (I) sample thermocouple, (J) base of detector, (K) mylar window, (L) gas chamber, (M) O-ring, (N) gas outlet, (O) cover of detector and (P) Mössbauer source connected to velocity transducer. Reprinted from Zhou and Morrish (1986). Copyright (1986). With permission from Elsevier.

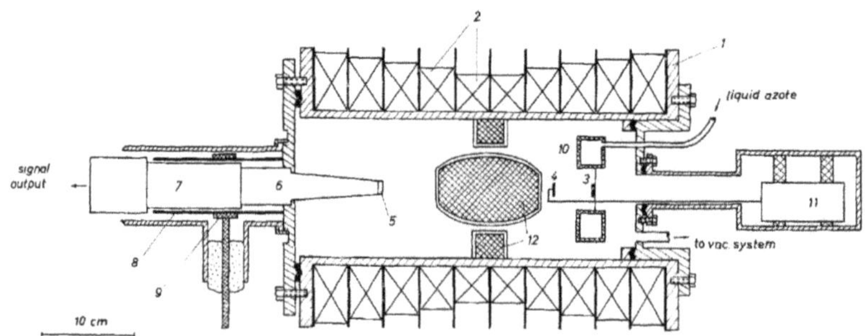

Figure 8.18. Diagram of a conversion electron Mössbauer spectrometer incorporating an energy selective electron detector, (1) body of the spectrometer, (2) coils, (3) sample, (4) Mössbauer source, (5) anthracene crystal, (6) lucite light guide, (7) photomultiplier tube, (8) mu-metal screen, (9) copper-ring with a rod dipped in liquid nitrogen for the photocathode cooling, (10) ring-cryostat for the sample, (11) velocity transducer and (12) lead screens. Reprinted from Bonchev *et al* (1969). Copyright (1969). With permission from Elsevier.

Bibliography and suggestions for further reading

Benczer-Koller N and Herber R H 1968 Experimental methods *Chemical Applications of the Mössbauer Effect* ed V I Goldanskii and R H Herber (New York: Academic) ch 2

Bonchev Z, Jordanov A and Minkova A 1969 Method of analysis of thin surface layers by the Mössbauer effect *Nucl. Instrum. Methods* **70** 36–40

Cook D C 2005 Spectroscopic identification of protective and non-protective corrosion coatings on steel structure in marine environments *Corros. Sci.* **47** 2550–70

Dunlap R A 2023 *An Introduction to the Physics of Nuclei and Particles* 2nd edn (Bristol: IOP Publishing)

Gifford W E 1966 The Gifford–McMahon cycle *Adv. Cryogenic Eng* **11** 152–9

Greenwood N N and Gibb T C 1971 *Mössbauer Spectroscopy* (London: Chapman and Hall)

Hesse J and Rübartsch A 1974 Model independent evaluation of overlapped Mössbauer spectra *J. Phys. E: Sci. Instrum.* **7** 526–32

Isozumi Y, Lee D I and Kádár I 1974 A new detector assembly for conversion electrons and x-rays from Mössbauer effect *Nucl. Instrum. Methods* **120** 23–8

Kohout P *et al* 2014 Optimized linear motor and digital PID controller setup used in Mössbauer spectrometer *AIP Conf. Proc.* **1622** 50–7

Kostroun V O and Crasemann B 1968 Total internal conversion coefficient of the 23.9-keV ground-state transition in ^{119m}Sn *Phys. Rev.* **174** 1535–9

Kovats T A and Walker J C 1969 Mössbauer absorption in Fe^{57} in metallic iron from the Curie point to the γ–δ transition *Phys. Rev.* **181** 610–8

Le Caër G and Dubois J M 1979 Evaluation of hyperfine parameter distributions from overlapped Mössbauer spectra of amorphous alloys *J. Phys. E: Sci. Instrum.* **12** 1083–90

Maksimova A A and Oshtrakh M I 2021 Applications of Mössbauer spectroscopy in meteoritical and planetary science, part I: undifferentiated meteorites *Minerals* **11** 612

Rancourt D G and Ping J Y 1991 Voigt-based methods for arbitrary-shape static hyperfine parameter distributions in Mössbauer spectroscopy *Nucl. Instrum. Meth.* B **58** 85–97

Rubinson W and Gopinathan K P 1968 Conversion coefficients of the Fe^{57} 14.4-keV transition, the K-capture fraction in Co^{57} e^- capture, and the K-fluorescence yield of Fe *Phys. Rev.* **170** 969–78

Salvat F and Parellada J 1984 Theory of conversion electron Mössbauer spectroscopy (CEMS) *Nucl. Instrum. Methods Phys. Res.* **1** 70–84

Senthil Kumar M, Reddy K V and Rama Rao K V S 1996 A low-temperature facility for Mössbauer and electrical resistivity studies using a closed-cycle helium refrigerator *Meas. Sci. Technol.* **7** 965–8

Shenoy G K and Wagner F E (ed) 1978 *Mössbauer Isomer Shifts* (Amsterdam: North-Holland)

Shigematsu T, Pfannes H D and Keune W 1980 Depth-selective conversion-electron Mössbauer spectroscopy *Phys. Rev. Lett.* **45** 1206–9

Stevens J G and Gettys W L 1978 Mössbauer isomer shift reference scales *Mössbauer Isomer Shifts* ed G K Shenoy and F E Wagner (Amsterdam: North-Holland) pp 901–6

Wang C, Hanrahan T and Johnson M 2018 High capacity closed-cycle 1 K cryocooler *Cryogenics* **95** 64–8

Window B 1971 Hyperfine field distributions from Mössbauer spectra *J. Phys. E: Sci. Instrum.* **4** 401–2

Zhou X and Morrish A H 1986 Simple high-temperature proportional detector for conversion-electron Mössbauer spectroscopy *Nucl. Instrum. Methods Phys. Res.* **18** 215–9

Richard A Dunlap

Chapter 9

Applications of Mössbauer spectroscopy to physics, chemistry and materials science

9.1 Introduction

The ability to observe Mössbauer spectra from the 14.4 keV transition in ^{57}Fe, has been a significant factor in the application of Mössbauer effect to the investigation of materials. Iron is very common and is an important element in many research and commercial fields. It has applications as a structural material and as a component in many magnetic materials. Since Mössbauer's initial discovery, well over 100 000 scientific publications on Mössbauer effect spectroscopy have appeared in the literature. About two thirds of these deal with ^{57}Fe. However, there have been substantial and important applications of other Mössbauer nuclides. About 18% of published reports deal with ^{119}Sn and the remaining 15% or so deal with about 43 other elements. In the present chapter we overview a few (of the very many) applications of the Mössbauer effect dealing with physics, chemistry and materials science.

9.2 General relativity

Although the majority of applications of Mössbauer effect spectroscopy deal with the properties of materials and are, therefore, of interest to solid state physicists, chemists, materials scientists or mineralogists, as discussed below, one of the earliest applications of the Mössbauer effect dealt with Einstein's theory of general relativity. In 1959 Robert Pound (1919–2010) and Glen Rebka, Jr (1931–2015) at Harvard University used the Mössbauer effect to provide the first laboratory demonstration of the validity of Einstein's theory of general relativity (Pound and Rebka 1959). Previous proofs of validity were based on astronomical observations. This experiment measured the gravitational redshift of photons (the Mössbauer gamma rays) which is equivalent to Einstein's hypothesis that clocks in different gravitational potentials run at different rates.

doi:10.1088/978-0-7503-6039-5ch9

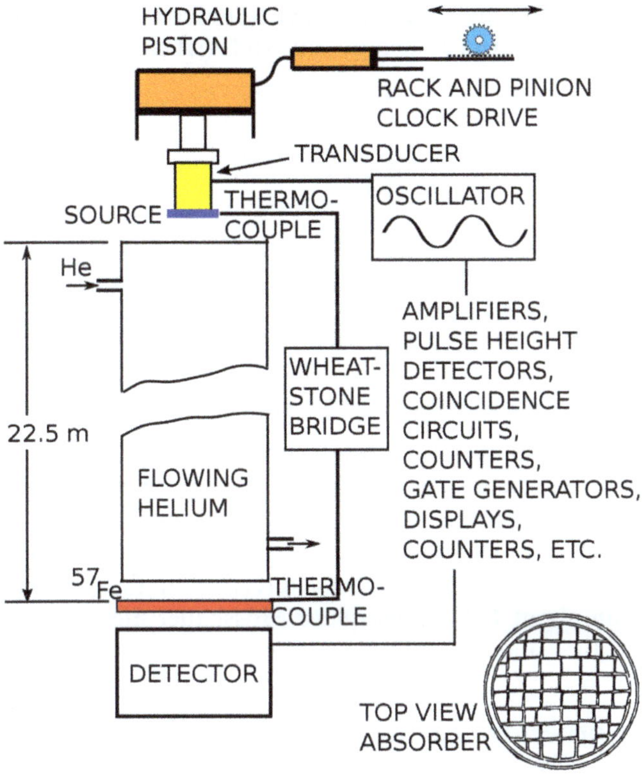

Figure 9.1. Schematic diagram of the experimental arrangement used by Pound and Rebka (1959) to observe the gravitational redshift using the Mössbauer effect. This [Pound and Rebka experimental arrangement] image has been obtained by the author from the Wikimedia website where it was made available by Prokaryotic Caspase Homolog (2023) under a CC BY-SA 4.0 licence. It is included within this chapter on that basis. It is attributed to Prokaryotic Caspase Homolog.

The experiment uses a source and absorber which are configured vertically so that there is a difference in the earth's gravitational potential between them, as illustrated in figure 9.1. This gravitational field will shift the ground and excited state energy levels differently and this difference will be observed in the Mössbauer spectrum (analogous to the chemical isomer shift). If the vertical relationship between the source and the absorber are reversed the difference will be in the opposite direction (in Doppler velocity). Specifically, when the source is placed in a higher position than the absorber, then the gamma rays from the source will be blue shifted, while when the source is in a lower position, the gamma rays will be red shifted. In order to optimize the magnitude of this effect, the vertical distance between the source and absorber should be maximized. This made the experiment particularly difficult because of the $1/r^2$ decrease in the intensity of the radiation from the source.

Pound and Rebka's experiment was performed with a vertical distance between the source and absorber of 22.5 m, as illustrated in figure 9.1. The gamma rays travelled through a helium-filled tube in order to reduce scattering of the photons in air. General relativity predicts an energy blue shift (when the source is at the top) of

$$\Delta E = E_\gamma \left[\sqrt{\frac{1 - \frac{2GM}{(R+h)c^2}}{1 - \frac{2GM}{Rc^2}}} - 1 \right]$$

where G is the gravitational constant, M is the mass of the earth, R is the earth's radius and h is the height of the source above the absorber. This energy shift is observed as a Doppler velocity as

$$v = \frac{\Delta E c}{E_\gamma}.$$

The difference in the Doppler shift for the two source/absorber configurations is predicted, by Einstein's theory, to be 1.5×10^{-3} mm s^{-1}. This Doppler velocity is less than the uncertainty in typical isomer shift measurements, which is about 0.005 mm s^{-1}, and indicates the difficulty of the measurement. The results obtained in the original Harvard University experiments agreed with the predictions of general relativity to within about 10%. Subsequent Mössbauer measurements of the same type have proved the validity of Einstein's theory to within about 1%.

9.3 Magnetic ordering studies

It can be shown that in a ferromagnet the hyperfine magnetic field scales as the spontaneous bulk magnetization, as shown in figure 6.5. In an antiferromagnet the hyperfine magnetic field scales as the spontaneous staggered (or sublattice) magnetization. While the measurement of the bulk magnetization of a ferromagnet is straightforward, it is not possible using bulk techniques, to observe the staggered magnetization of an antiferromagnet. This is because, as described above, the net magnetization is zero in the magnetically ordered state. The Mössbauer effect, however, is a simple way of measuring the hyperfine magnetic field. The hyperfine field arises because only the near neighbors (first and maybe the second nearest neighbors) of the Mössbauer atom influence the hyperfine field, so that there is no large-scale spatial vector sum that averages to zero. The sensitivity of the Mössbauer effect to the very local environment around the probe nucleus is one of this technique's strengths and, as discussed further below, gives it the ability to nicely complement other techniques (e.g., x-ray diffraction and magnetic susceptibility measurements).

Figure 9.2 shows the ^{57}Fe Mössbauer effect spectra of a common antiferromagnet, FeF_3 in the vicinity of the Néel temperature. As can be seen in the figure, above T_N, where the material is paramagnetic, the spectrum consists of a single absorption line. This indicates the lack of magnetic Zeeman splitting and also indicates the lack of an electric quadrupole interaction. This latter feature results from the high degree of symmetry of the octahedrally coordinated iron sites. Below the Néel temperature the spectra show magnetically split sextets as a result of the hyperfine magnetic field

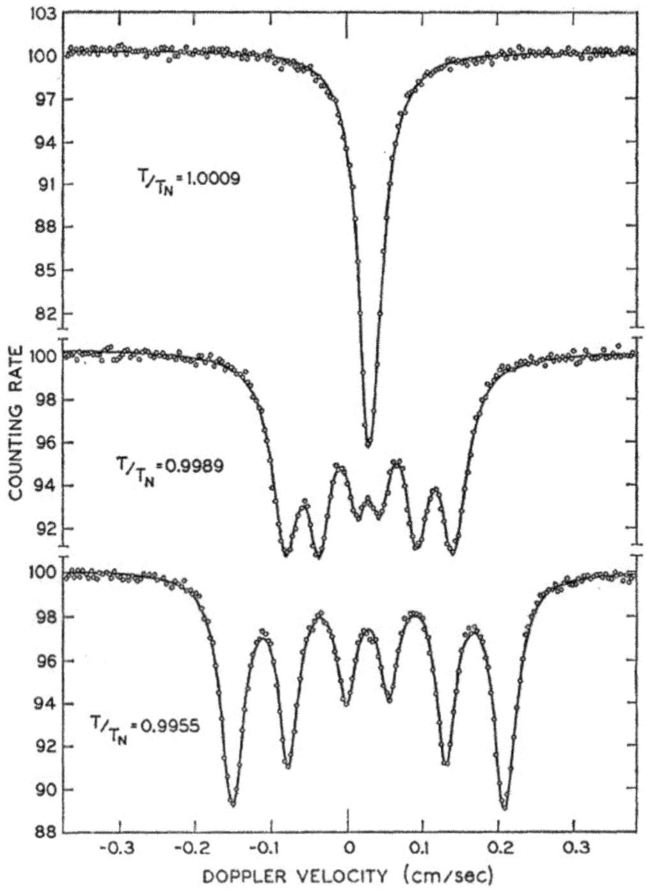

Figure 9.2. ^{57}Fe Mössbauer effect spectra of FeF_3 as a function of temperature above and below the Néel temperature, $T_N = 363$ K. Reprinted with permission from Wertheim *et al* (1968). Copyright 1968 by the American Physical Society.

present at the iron nuclei which result from the magnetic order. As clearly seen in the figure, the magnetic splitting increases as the temperature decreases as a result of the increasing hyperfine magnetic field. This corresponds to the increasing spontaneous staggered magnetization as the thermal energy is decreased.

The temperature dependence of the hyperfine magnetic field can be determined from an analysis of the spectra following from equations (7.3) and (7.4). The measured temperature dependence of the hyperfine magnetic field in FeF_3 is illustrated in figure 9.3. This temperature dependence shows the same general trend as the magnetization of a ferromagnet illustrated in figure 6.5. While mean field theory shows rough agreement with experimental results, the analysis of data such as that in figure 9.3, allows for the validation of more detailed models of magnetic interactions in solids.

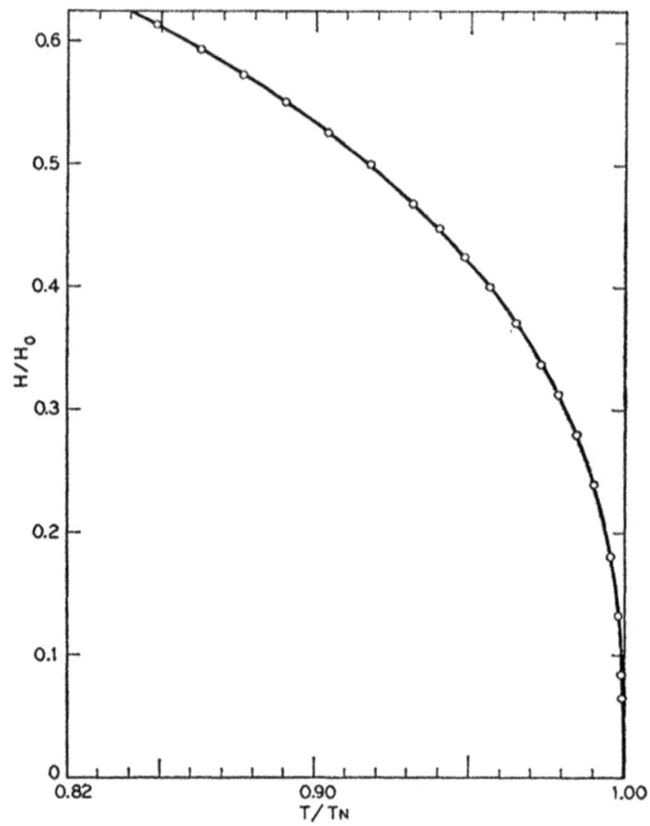

Figure 9.3. Temperature dependence of the hyperfine magnetic field at the iron nucleus in FeF_3 in the vicinity of the Néel temperature. Reprinted with permission from Wertheim *et al* (1968). Copyright 1968 by the American Physical Society.

9.4 Crystallographic structure studies

Magnetite, Fe_3O_4, as discussed previously, provides an interesting example of how Mössbauer spectroscopy can further our understanding of the structural properties of materials and how particular magnetic properties are related to crystal structure.

Figure 9.4 shows a room temperature ^{57}Fe Mössbauer effect spectrum of magnetite (Fe_3O_4) with a small amount of Co impurity. This spectrum clearly shows two Zeeman split sextets. The sextets have different isomer shifts and different splittings giving rise to the feature seen in the figure that the lines of the two sextets are separated on the left-hand (negative velocity) side of the spectrum but overlap with each other on the right-hand (positive velocity) side of the spectrum. The parameters obtained from the analysis of this spectrum are summarized in table 9.1. These results can be considered in the context of the previous discussion of the magnetite structure. The tetrahedrally coordinated iron atoms are all valence 3+ and are responsible for the component of the spectrum with the less positive isomer shift and the larger hyperfine magnetic field. The octahedrally coordinated iron atoms are

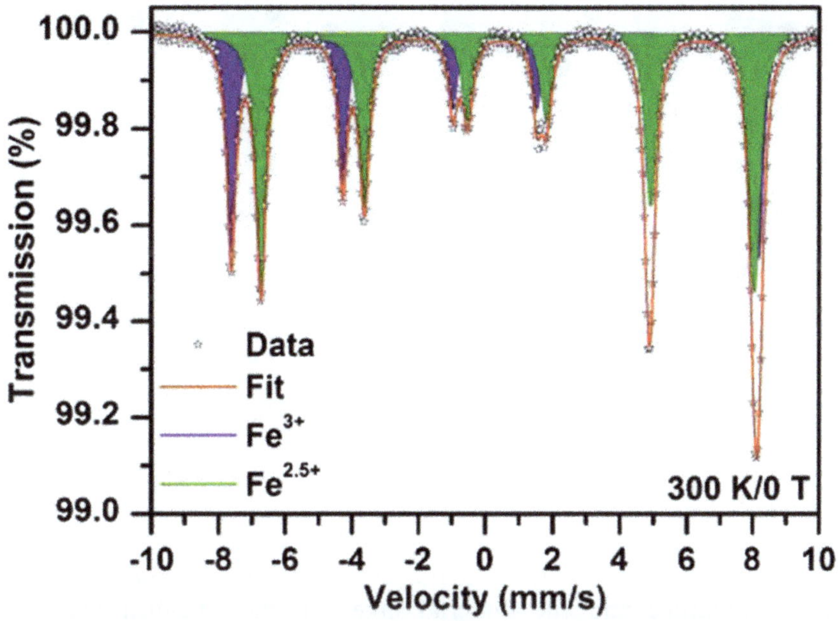

Figure 9.4. Room temperature ^{57}Fe Mössbauer effect spectrum of magnetite ($Co_{0.01}Fe_{2.99}O_4$), showing the tetrahedral Fe^{3+} (A-sites) and octahedral Fe^{2+}/Fe^{3+} (B-sites) components of the spectrum in green and purple, respectively. Reprinted by permission from Datta *et al* (2017) CC BY 4.0.

Table 9.1. Room temperature ^{57}Fe Mössbauer effect parameters for Fe_3O_4 from a fit to two sextets. Isomer shifts are relative to room temperature α-Fe.

Site	Valence	Isomer shift (mm s^{-1})	Hyperfine field (Tesla)
Tetrahedral	Fe^{3+}	+0.28	49.1
Octahedral	Fe^{2+}/Fe^{3+}	+0.66	46.0

half 3+ and half 2+ valence state. These two groups of octahedrally coordinated iron atoms occupy crystallographically equivalent sites. At room temperature, one of the valence electrons can hop between atoms on a time scale which is less than the characteristic Mössbauer measurement time (i.e., the lifetime of the 14.4 keV excited state). Thus, the octahedral 2+/3+ spectra cannot be distinguished, leading to an isomer shift for the octahedral iron which is intermediate between typical Fe^{2+} and Fe^{3+} isomer shifts (compare table 9.1 with figure 5.6).

Another interesting application of the Mössbauer effect deals with site occupancy of disordered iron alloys. Fe–Ga is a good example. Pure iron has the bcc structure and Ga is readily soluble in iron. For small amounts of Ga, the Ga atoms will substitute randomly for the iron atoms. For larger amounts of Ga, the Ga atoms may exhibit preferential site occupancy. The exact distribution of Ga atoms depends on the Ga concentration as well as the method by which the sample has been

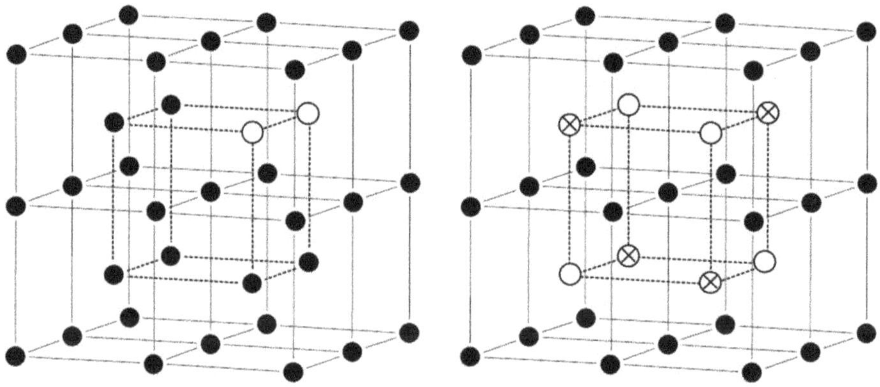

Figure 9.5. Crystal structures of α-Fe with substitutional Ga impurities with (a) impurities located at random lattice sites and (b) the $D0_3$ structure of Fe_3Ga with an ordered Ga sublattice. Reprinted from Dunlap *et al* (2006). Copyright (2003). With permission from Elsevier.

prepared. The properties of Fe–Ga are particularly interesting when the Ga concentration approaches 25 at%. At this concentration the Ga atoms can form an ordered sublattice and the material takes on the so-called $D0_3$ structure. Figure 9.5 shows eight bcc iron unit cells and illustrates some possible structures. In figure 9.5(a) the Ga atoms substitute randomly for iron atoms. In figure 9.5(b) the Ga atoms in Fe_3Ga form an ordered sublattice by occupying alternating body center sites. It is found, as described below, that this preferential site occupancy can begin at around 18 at% Ga.

Since the magnetic properties of Fe–Ga vary as a function of composition and sample preparation method, it is important to understand the relationship between these properties and the structural ordering. Thus, we need a suitable experimental method for investigating the structure and being able to distinguish between the situations in figures 9.5(a) and (b). X-ray diffraction patterns of two Fe–Ga alloys, one with 17 at% Ga and one with 19 at% Ga are shown in figure 9.6. These two patterns appear virtually identical. While x-ray diffraction can, in principle, provide information about the distribution of atoms through an analysis of the relative intensity of the various diffraction peaks, the technique is not effective in this particular case. The sensitivity of x-ray diffraction to chemical order, depends on its ability to distinguish the scattering from the different atomic species in the material. The scattering factor for a particular element depends on its number of electrons. For this particular alloy, an iron atom has 26 electrons, while a gallium atom has 31 electrons, and this difference is not large enough to allow for the subtleties of different Ga site occupancies to be distinguished.

The Mössbauer effect, however, is much more sensitive to this type of chemical order. This is because the Mössbauer effect is a very local probe of the structural and chemical order and, from a magnetic standpoint, an iron atom looks very different than a Ga atom. Therefore, from the viewpoint of the Mössbauer effect, the distribution of local environments around the iron atoms in figure 9.5(a) is quite different than it is in figure 9.5(b).

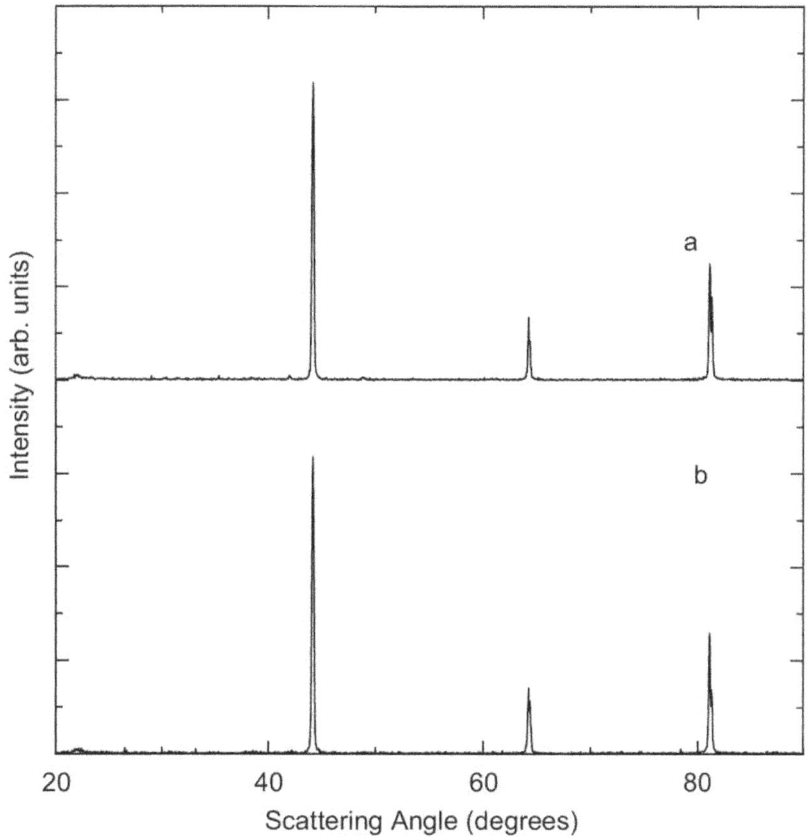

Figure 9.6. X-ray diffraction patterns of (a) $Fe_{83}Ga_{17}$ and (b) $Fe_{81}Ga_{19}$. Reprinted from Gaudet *et al* (2008). Copyright (2008). With permission from Elsevier.

Figure 9.7 shows the room temperature ^{57}Fe Mössbauer effect spectra for the two samples as in figure 9.6. Even a simple visual inspection shows that these two spectra are quite different. The spectrum of the 19 at% Ga sample is substantially more complex than the spectrum of the 17 at% Ga sample. These spectra were fit to Voigt-based hyperfine field distributions as discussed in the previous chapter. The analysis of the spectra in figure 9.7 demonstrates that the 17 at% Ga spectrum consists of a relatively featureless broad distribution of hyperfine fields, as shown in the right panel of the figure. This indicates that the iron atoms have a rather random distribution of sites with, on the average, 17% Ga nearest neighbors. The hyperfine magnetic field distribution of the 19 at% Ga sample shows the presence of three distinct iron site fields corresponding to three different, and relatively well defined, crystallographic environments. These sites have mean fields of 20.7, 25.8 and 31.6 T. If we look in detail at the iron and gallium ordering in the $D0_3$ structure, as shown in figure 9.5(b), we see that there are two different iron sites, the $D0_3$ A-sites, where the iron is at the center of a cube with four iron and four gallium nearest neighbors, and the $D0_3$ D-sites, where the iron is at the center of a cube with eight iron nearest

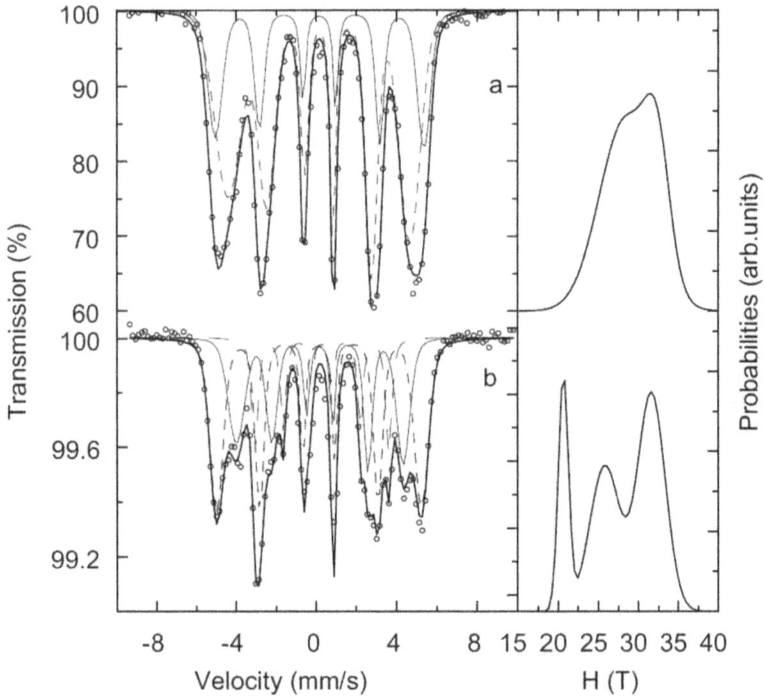

Figure 9.7. Room temperature ^{57}Fe Mössbauer effect spectra of (a) $Fe_{83}Ga_{17}$ and (b) $Fe_{81}Ga_{19}$. To the right the hyperfine field distributions as extracted from the Mössbauer spectra are shown. Reprinted from Gaudet *et al* (2008). Copyright (2008). With permission from Elsevier.

neighbors. These two sites correspond to iron nuclei with hyperfine magnetic fields of 20.7 and 31.6 T, respectively. Because the stoichiometry of the sample is only 19 at% Ga, not 25 at% Ga as in the ordered $D0_3$ structure, there are iron sites which have iron neighbors in locations where there should be gallium atoms. These sites give rise to the intermediate hyperfine magnetic field value of 25.8 T.

The example of Fe–Ga shows that Mössbauer effect spectroscopy is effective at determining chemical ordering on lattices in crystalline materials when more traditional techniques, such as x-ray diffraction, are less than ideal.

9.5 Hyperfine field distributions

Mössbauer effect spectra of materials with chemical and/or structural disorder exhibit distributed hyperfine parameters. This was illustrated in the previous section for crystalline alloys with chemical disorder. In the case of materials with structural disorder, there has been significant work reported in the literature dealing with distributed hyperfine Mössbauer parameters in amorphous materials. Non-magnetic amorphous materials exhibit a distribution of quadrupole splittings (along with a distribution of isomer shifts) and this situation is described in detail in section 10.3. Magnetic amorphous materials exhibit a distribution of hyperfine fields (along with a distribution of isomer shifts) and this situation is described in the present section.

Here we consider two examples of distributed hyperfine fields as studied by Mössbauer spectroscopy in amorphous metals. Firstly, the investigation of Fe- and Sn-containing alloys using both ^{57}Fe Mössbauer effect and ^{119}Sn Mössbauer effect. Secondly, the effects of neutron irradiation on the Fe hyperfine field distribution in Fe-based metallic glasses.

9.5.1 Studies of Fe- and Sn-containing alloys

Amorphous alloys containing both Fe and Sn have been prepared by melt spinning and have been investigated by Dunlap (1982, 1984) and Miglierini *et al* (2020). Figure 9.8 shows ^{57}Fe Mössbauer effect spectra of amorphous $Fe_{78}B_{15}Sn_7$ as reported by Miglierini *et al* (2020) at room temperature and at 4.2 K (both with and without an applied external field). Figures 9.8(a) and (b) show that the field increases as the temperature is decreased, consistent with expectations for the temperature dependence of the magnetization (see section 9.3). The hyperfine field

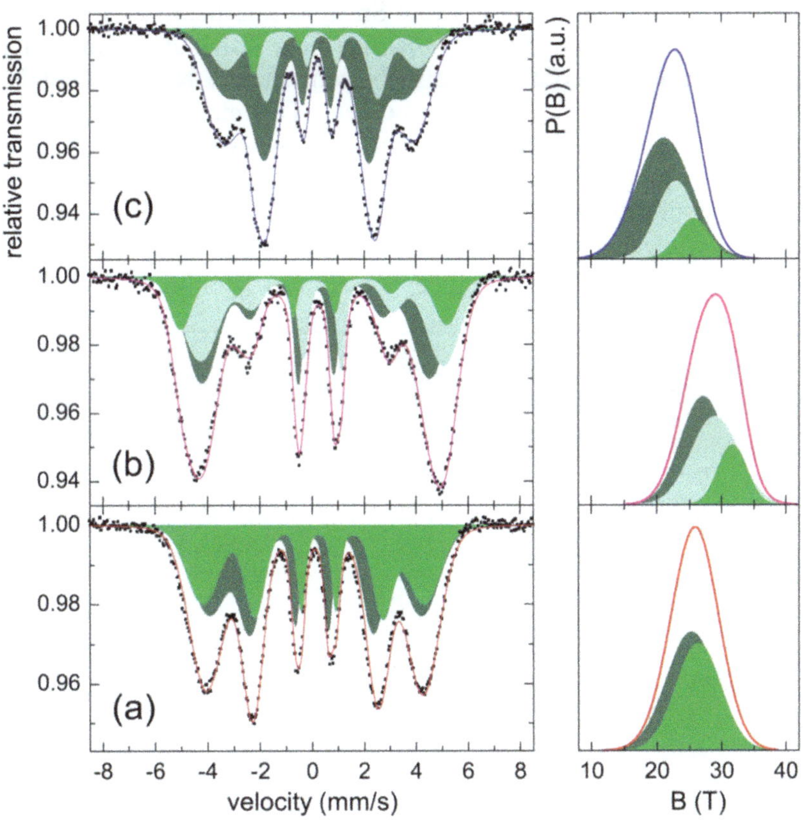

Figure 9.8. (left panel) ^{57}Fe Mössbauer effect spectra of amorphous $Fe_{78}B_{15}Sn_7$ in the as-quenched state (a) at 300 K, (b) at 4.2 K and (c) at 4.2 K in an external magnetic field of 6 T in the plane of the absorber. (right panel) Corresponding hyperfine field distributions for the ^{57}Fe Mössbauer spectra in the left panel. Reprinted from Miglierini *et al* (2020). Copyright (2020). With permission from Elsevier.

distribution in an external magnetic field is shifted to lower values indicating that the internal field at the iron nuclei is oriented antiparallel to the external field. That is, the nuclear hyperfine field is antiparallel to the Fe atomic magnetic moment and in this case the internal field is referred to as negative. This is consistent with the internal field orientation in α-Fe (see Reissner 2021). At room temperature the relative line intensities give an angle $\theta = 58°$ between the magnetization direction and the γ-ray direction (see section 7.4), which is close to the magic angle for a random orientation of magnetic moments. At 4.2 K, the angle is $\theta = 33°$, indicating a net magnetization vector that is rotated out of the plane of the sample. This behavior is characteristic of temperature dependent stress induced magnetic anisotropy. In an external magnetic field, $\theta = 77°$ and this indicates that the applied field has partially rotated the net magnetization back into the sample plane (i.e., the direction of the applied field).

The presence of Sn in samples such as those described above provides the opportunity of performing ^{119}Sn Mössbauer studies in addition to ^{57}Fe Mössbauer studies. Figure 9.9 shows some results for some Sn-containing Fe-based amorphous alloys, from Dunlap (1984). These show Sn hyperfine field distributions that are fundamentally different than the Fe hyperfine field distributions in similar amorphous alloys. The Fe hyperfine field distributions show a roughly Gaussian shape that has zero probability of zero hyperfine field. This means that all Fe sites have hyperfine fields that are of the same sign, i.e., negative, or antiparallel to an externally applied magnetic field. The Sn hyperfine field distributions, on the other hand, show a significant probability of zero internal field at the Sn nuclei. Since a single Mössbauer measurement is not sensitive to the sign of the internal field, this

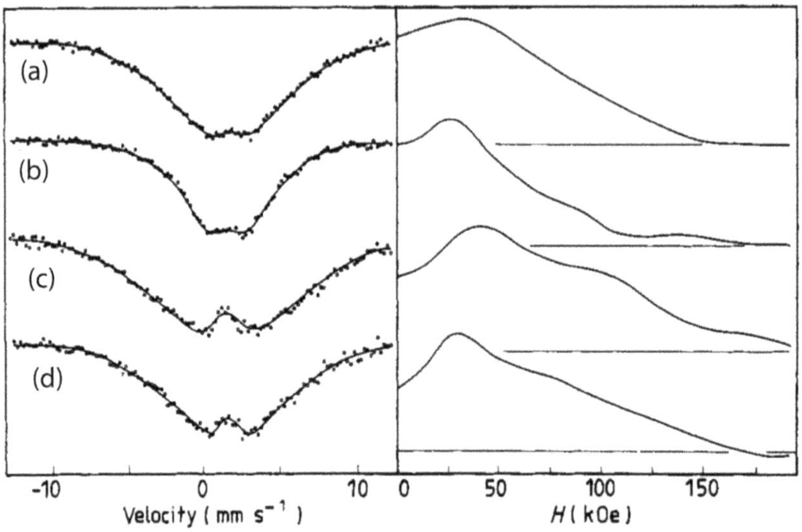

Figure 9.9. Room temperature ^{119}Sn Mössbauer spectra and hyperfine field distributions for the amorphous alloys (a) $Fe_{79}SnB_{20}$, (b) $Fe_{69}SnCr_{10}B_{20}$ (c) $Fe_{40}Ni_{39}SnB_{20}$ and (d) $Fe_{35}Ni_{34}SnCr_{10}B_{20}$. Note: 1.0 kOe = 0.1 T. From Dunlap (1984) copyright IOP Publishing. Reproduced with permission. All rights reserved.

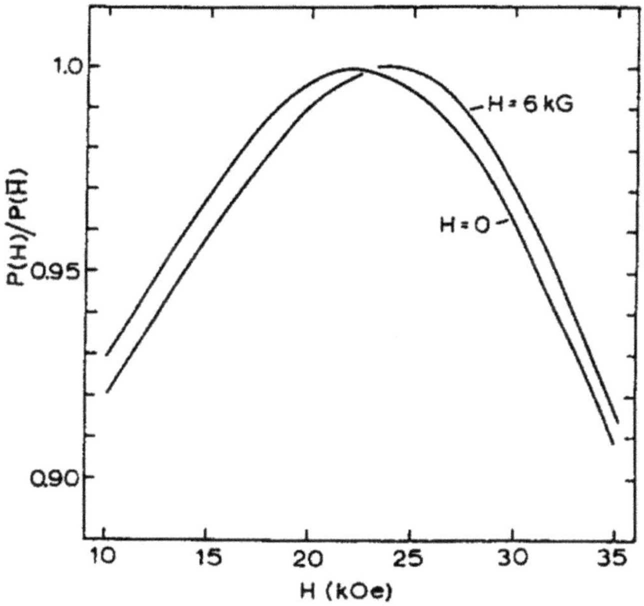

Figure 9.10. Details of the room temperature ^{57}Fe hyperfine field distribution near the most probable field for $Fe_{78}Sn_2B_{20}$ with and without an applied field of approximately 0.6 T. Reprinted from Dunlap (1982). Copyright (1982). With permission from Elsevier.

means that Sn sites have fields of both signs and the horizontal axis on the field distribution plots in figure 9.9 actually represent the absolute value of the field where distribution components of one sign are folded onto those of the opposite sign. This feature can be understood in more detail by conducting Sn Mössbauer effect measurements in an externally applied magnetic field. Results of such an investigation are illustrated in figure 9.10. These results indicate a peak Sn field of about 2.5 T and that the application of an external magnetic field shifts the peak in the Sn hyperfine field distribution to a higher field. That is, the internal Sn site field aligns parallel to the external field and the mean Sn field is positive.

The results given above for Sn hyperfine fields in Fe-based amorphous alloys can be understood quantitatively by looking at the results of Sn Mössbauer studies of dilute Sn impurities in crystalline α-Fe. These measurements give an internal field at the Sn site of −8.0 T (see Campbell and Blandin (1975)) compared to about +2.5 T in the amorphous Fe-based alloys. It is important to note that Sn atoms do not carry a magnetic moment, so the field that is observed is the result of the moments associated with the neighboring Fe atoms. It has been suggested (Campbell and Blandin 1975) that the transferred field at the Sn site in a ferromagnetic host can be expressed in terms of the magnetic moments on the first and second nearest neighbors, μ_1 and μ_2, respectively, as

$$H(\mathrm{Sn}) = -14\mu_1 + 10\mu_2 \tag{9.1}$$

where H is in T and μ_1 and μ_2 are in μ_B. This indicates that the first neighbor magnetic moments provide a negative contribution to the Sn hyperfine field and the second neighbor magnetic moments provide a positive contribution. For Sn in crystalline α-Fe, $\mu_1 = \mu_2 = 2.2\ \mu_B$, and equation (9.1) yields reasonable agreement with the measured Sn hyperfine field. In an amorphous alloy, the there are no well-defined near neighbor distances for the Sn sites and variations in the relative importance of the positive and negative contributions in equation (9.1) will lead to a distribution of hyperfine fields, as observed experimentally. The results obtained, as in figure 9.9 indicate that, contrary to the situation for crystalline α-Fe, the Sn hyperfine field in amorphous Fe-based alloys is dominated by the positive contribution.

9.5.2 Studies of the effects of neutron irradiation

Mössbauer effect spectroscopy is an effective means of investigating the microstructure of solids and, in particular, has been important, as noted above, in understanding the distribution of environments in amorphous materials. Recent work by Miglierini (2021) has shown the effects of neutron irradiation on the microstructure of amorphous alloys by the application of Mössbauer spectroscopy. The effects of neutron irradiation on alloys containing both boron and chromium are of particular intertest. This is because of their potential use as shielding material for neutrons emitted by fission reactors (see Khong *et al* 2016). Boron (specifically ^{10}B, which is 20% naturally abundant) has a very large thermal neutron absorption cross-section. The ^{10}B undergoes the reaction,

$$n + {}^{10}B \rightarrow {}^{7}Li + \gamma.$$

The α-particles (at 1.4 MeV) and the ^{7}Li ion (at 0.9 MeV), are both high energy positive ions which can cause significant structural damage in the alloy but are readily stopped within the material. The chromium content in the alloy is advantageous as it improves corrosion resistance.

Figure 9.11 shows room temperature ^{57}Fe Mössbauer effect spectra of amorphous $Fe_{80}Cr_2Si_{14}B_{14}$ and their corresponding hyperfine field distributions (Miglierini 2021). Results are shown for non-irradiated and neutron irradiated samples. Microstructural effects caused by neutron irradiation are seen to produce measurable changes in the Mössbauer spectra. Irradiation causes an increase in the high field component of the hyperfine field distribution as evidenced by the negative value shown for the difference spectrum in figure 9.11(c). Irradiation also causes a decrease in the intensity of lines 2 and 5 of the spectrum, indicating a rotation of the magnetization axis towards the out-of-plane direction.

Additional information concerning the effects of neutron irradiation of the microstructure of amorphous metals can be obtained by looking at the effects of chromium content on the results of Mössbauer investigations of neutron irradiated materials. It is known that the addition of chromium to amorphous Fe–B-based alloys has two important effects (see e.g., Dunlap 1984 and references therein). Firstly, the presence of chromium drastically reduces the Curie temperature of the

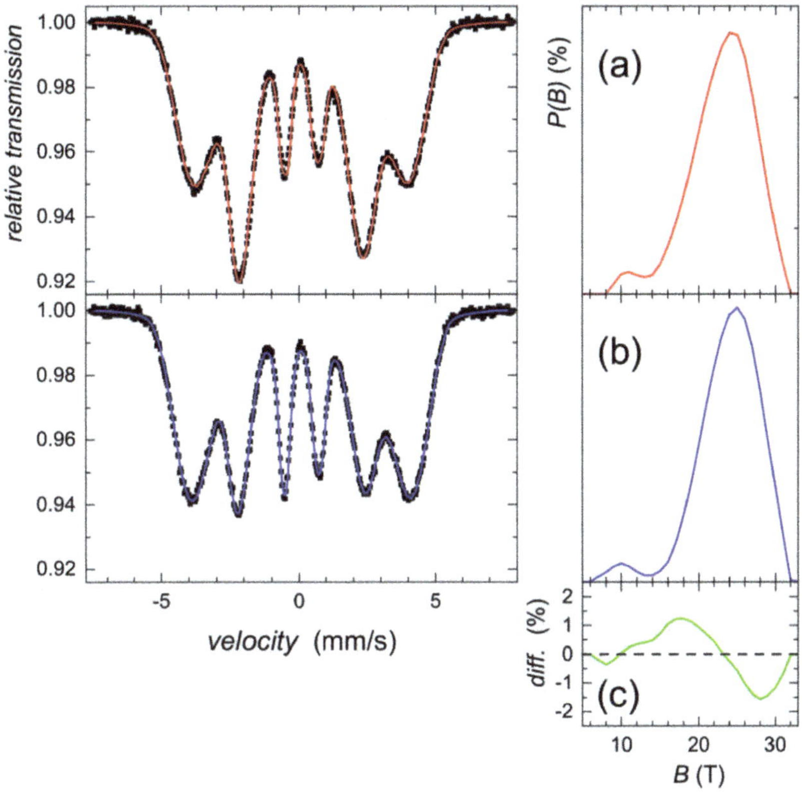

Figure 9.11. (Left) Room temperature ^{57}Fe Mössbauer spectra of amorphous $Fe_{80}Cr_2Si_{14}B_{14}$ and (right) corresponding hyperfine field probability distributions. (a) Non-irradiated sample, (b) sample after neutron irradiation with a fluence of 10^{19} n cm^{-2} and (c) differences between non-irradiated and irradiated distributions. From Miglierini (2021). Copyright 2021 by the author. CC BY 4.0

alloy and secondly, the hyperfine field distribution becomes clearly bimodal as a result of the formation of two distinctly different Fe environments due to short-ranged chromium clustering. Results for non-irradiated and irradiated alloys as a function of chromium content from Miglierini (2021) are shown in figure 9.12. It is clear in figure 9.12(a) that the effects that are characteristic of chromium substitutions in the non-irradiated Fe-based amorphous alloys are observed. The hyperfine field distribution progressively shifts to lower field values. As well, the bimodal nature of the distribution becomes more obvious with the low field component due to Fe sites near chromium clusters becoming more prominent. The effects of neutron irradiation are shown in figure 9.12(b), where it is clear that the decrease in the mean hyperfine field as a function of increasing chromium content is more rapid than in the non-irradiated samples. This behavior results from the fact that minor changes in microstructure can have a pronounced effect on the resulting magnetic properties when the system is close to a critical transition. Thus, when the addition of chromium to amorphous Fe-based alloys lowers the Curie

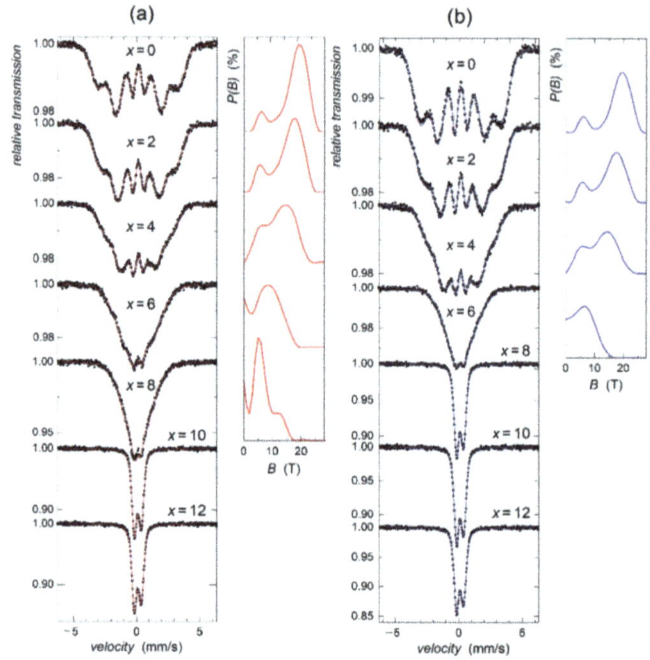

Figure 9.12. Room temperature ^{57}Fe Mössbauer spectra and corresponding hyperfine field probability distributions of amorphous $Fe_{30}Ni_{48-x}Cr_xMo_2Si_5B_{15}$ alloys as a function of chromium content. (a) Non-irradiated samples and (b) samples after neutron irradiation with a fluence of 10^{19} n cm^{-2}. From Miglierini (2021). Copyright 2021 by the author. CC BY 4.0

temperature close to room temperature, then hyperfine field distributions at this temperature are more sensitive to neutron irradiation effects. These features are quantified in figure 9.13, where the normalized mean hyperfine field is plotted for alloys with different chromium content as a function of neutron fluence. It is clear that for chromium content of $x = 6$ and higher, there is a marked decrease in mean hyperfine field for fluences of 10^{18} n cm^{-2} and higher.

9.6 Impurity studies

Mössbauer effect impurity studies roughly fall into two categories, studies of dilute Mössbauer probe atoms in a host material and studies of dilute impurity atoms in a Mössbauer probe host. The number of possible systems that fall into these categories is very large, but as examples of what can be learned from such studies, we look at two classes of materials. Firstly, dilute iron in an elemental metallic host, M, that is alloys of the form, $\underline{M}$Fe, and secondly, dilute elemental metallic impurities in iron, that is alloys of the form, Fe$\underline{M}$.

9.6.1 $\underline{M}$Fe alloys

Mössbauer effect isomer shifts for dilute Fe impurities in various metallic hosts have been summarized in van der Woude and Sawatzky (1974) and Ingalls *et al* (1978). Figure 9.14

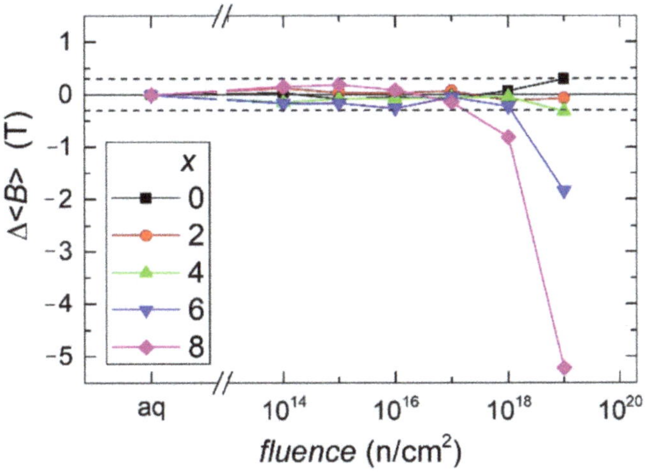

Figure 9.13. Change in the mean value of the room temperature ^{57}Fe hyperfine magnetic field as a function of neutron irradiation fluence and chromium content for the amorphous alloys $Fe_{30}Ni_{48-x}Cr_xMo_2Si_5B_{15}$. Values are referenced to the mean value for each composition in the non-irradiated state. From Miglierini (2021). Copyright 2021 by the author. CC BY 4.0

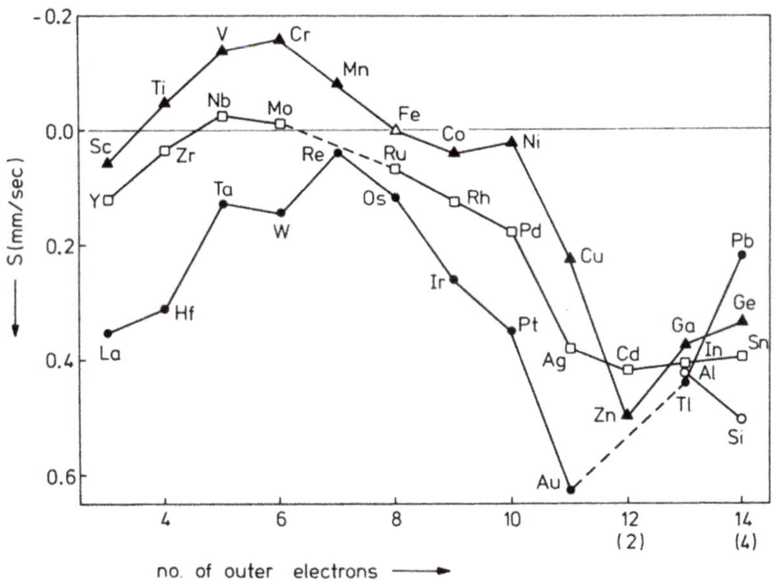

Figure 9.14. Measured room temperature ^{57}Fe isomer shifts for dilute Fe impurities in different elemental metallic hosts as a function of the number of outer electrons on the host atoms. Lines show results for the 3d, 4d and 5d series. Reprinted from Ingalls *et al* (1978). Copyright (1978). With permission from Elsevier.

shows measured ^{57}Fe isomer shifts as a function of the number of outer electrons on the host atoms for elements in the 3d, 4d and 5d series. The data in the figure show that for increasing number of d and outer sp electrons across series, the isomer shift first becomes less positive and after about six outer electrons, becomes more positive.

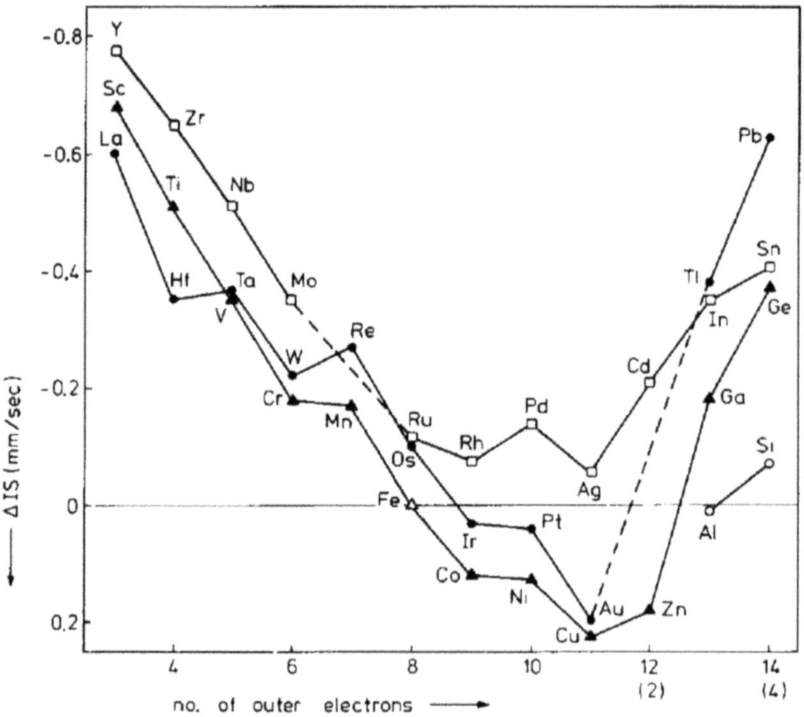

Figure 9.15. Measured change in room temperature ^{57}Fe isomer shifts for dilute Fe impurities in different elemental metallic hosts as a function of the number of outer electrons on the host atoms after correcting for atomic volume effects. Reprinted from van der Woude and Sawatzky (1974). Copyright (1974). With permission from Elsevier.

Following the discussion on Mössbauer isomer shifts in chapter 4, it is clear that the changes in the measured isomer shift of Fe impurities in a host material will depend on the electron density at the ^{57}Fe probe nuclei. The two most significant components to these changes as a function of the number of outer electrons on the host atoms are the atomic volume of the host atoms and charge transfer. Figure 9.15 shows the data from figure 9.14 that have been corrected for host atomic volume. These results should indicate the component of the charge density at the probe nuclei that results from charge transfer. The data in figure 9.15 show well defined trends. The isomer shift becomes increasingly positive as the relevant host atom d-shell is filled. After the d-shell is filled, the isomer shift becomes increasingly less positive with increasing number of outer sp electrons on the host atoms. On the basis of these trends, it can be concluded that non-transition metal hosts affect the electron density at the ^{57}Fe probe nuclei differently than transition metal hosts.

The general behavior shown in figure 9.15 can be described phenomenologically on the basis of the discussion of isomer shifts in section 4.3. We begin by considering 3d elements to the left of iron in the periodic table. These elements have fewer 3d electrons than iron and the isolated Fe impurity in this lattice tends to assume an electronic structure that is more like its neighbors by transferring 3d charge to the surrounding atoms. This reduction in localized 3d Fe electrons results in an increase in the 3s electron density at the Fe nucleus as a result of shielding effects. Analogous to the situation depicted in figure 4.6 for the isomer shift of iron in different valence states, decreasing the number of host d electrons leads to increasingly positive isomer shifts for the impurity ^{57}Fe probe atoms. For host elements to the right of Fe in the 3d series, the effects of the electronic configuration of the neighboring atoms lead to the opposite change in localized Fe d-electron density and a continuation of the isomer shift trends seen in figure 9.15.

The trend towards less positive isomer shifts after the d-band is filled can be explained by the increase in s-like character of the host outer electrons. The tendency of the Fe impurity atoms to take on the electronic characteristics of its neighbors means that the screening from s-like electrons gives rise to an increase in the electron density at the Fe nuclei.

The study of dilute AlFe alloys is also an interesting example of the ability of Mössbauer spectroscopy to understand the microstructure of solids. The equilibrium solubility of Fe in fcc aluminum is quite small (<0.01%), and this creates difficulty in performing ^{57}Fe Mössbauer effect experiments on this system, unless enriched ^{57}Fe is used in the absorber. Several studies, however, have shown that supersaturated fcc *Al*Fe alloys can be prepared with up to about 6 at% Fe by rapid quenching (see e.g., Nasu *et al* 1974, Dunlap *et al* 1988).

Figure 9.16 shows room temperature Mössbauer spectra from Nasu *et al* (1974) of splat quenched $Al_{1-x}Fe_x$ alloys. X-ray diffraction studies of these samples showed the presence of a single phase fcc structure, indicating that the alloys are supersaturated solutions of Fe in Al. Similar results have been reported by Dunlap *et al* (1988) for roller quenched samples of $Al_{1-x}Fe_x$.

The spectra in figure 9.16 have been fit to two components, a singlet with an isomer shift at +0.424 mm s^{-1} relative to room temperature α-Fe and a doublet with an isomer shift of about +0.15 mm s^{-1} and a quadrupole splitting of about 0.35 mm s^{-1}. The singlet results from isolated Fe probe atoms in the fcc aluminum host matrix. These have a symmetric 12-coordinated local environment and do not show any quadrupole splitting from an electric field gradient. The measured isomer shift is consistent with that from other measurements of very dilute Fe Mössbauer probe atoms in an aluminum host, as shown in figure 9.14. The doublet results from Fe probe atom with one or more first neighbor Fe atoms. This arrangement makes the probe atom more Fe-like and yields a less positive isomer shift than for the isolated Fe probe atoms. It also gives rise to a local asymmetry round the probe atom and a corresponding quadrupole splitting.

The probability of an impurity probe atom local environment corresponding to the singlet and doublet can be readily calculated for a random distribution of impurity atoms. The probability of a probe atom having n impurity nearest neighbors is given in terms of the impurity concentration, x, and the local coordination number, c, by the binomial coefficient as

$$P(c, n, x) = \binom{c}{n} x^n (i - x)^{c-n} = \frac{c!}{(c - n)!n!} x^n (i - x)^{c-n}. \tag{9.2}$$

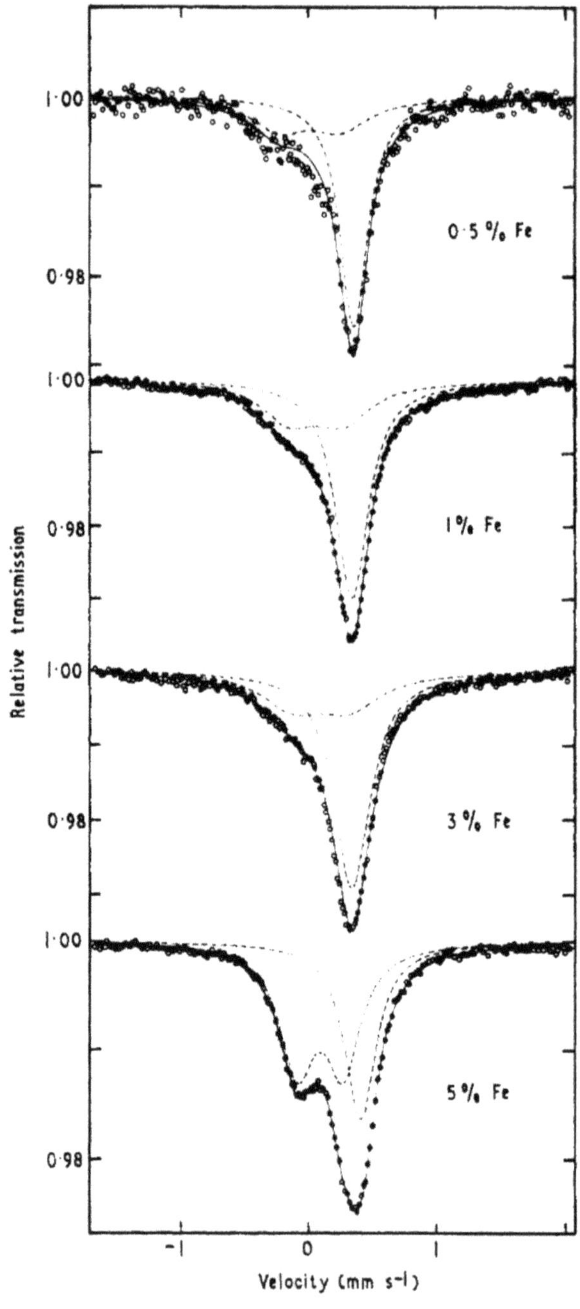

Figure 9.16. Room temperature ^{57}Fe Mössbauer effect spectra of supersaturated $\underline{Al}$Fe alloys. From Nasu *et al* (1974) copyright IOP Publishing. Reproduced with permission. All rights reserved.

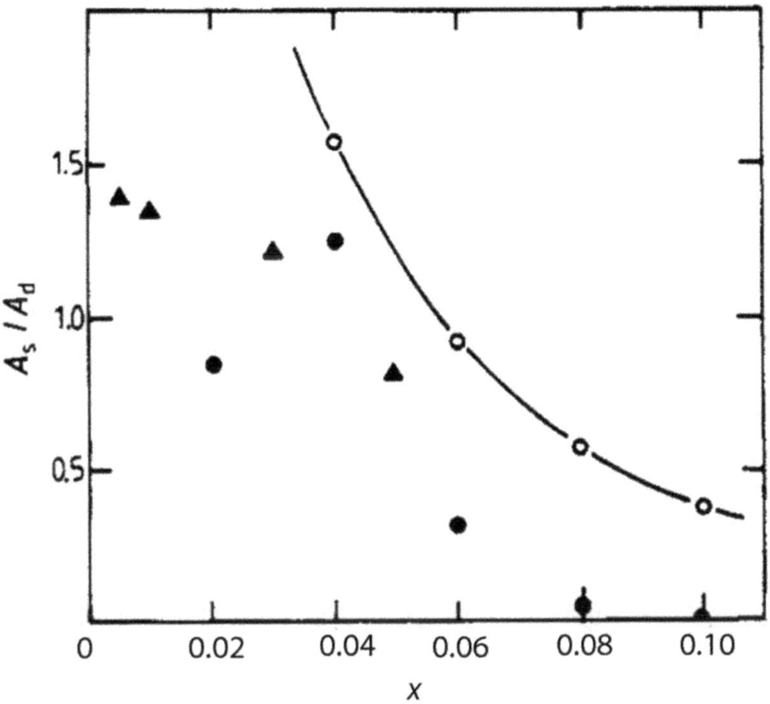

Figure 9.17. Relative areas of the singlet and doublet for Mössbauer spectra of dilute AlFe alloys as a function of Fe concentration. Experimental results from Nasu *et al* (1974) are shown by the solid triangles (▲), experimental results from Dunlap *et al* (1988) are shown by solid circles (●) and calculated results from equations (9.2) and (9.3) are shown by the open circles (○) and the solid line. From Dunlap *et al* (1988)

The relative areas of the singlet, A_s, and doublet, A_d, in the Mössbauer spectrum can be expressed as

$$\frac{A_s}{A_d} = \frac{P(c,\,0,\,x)}{1 - P(c,\,0,\,x)}. \tag{9.3}$$

Results for A_s/A_d from Nasu *et al* (1974) and Dunlap *et al* (1988) are shown in figure 9.17 as a function of x along with the theoretical values as calculated from equations (9.2) and (9.3).

It is clear that the results from Nasu *et al* (1974) and Dunlap *et al* (1988), are reasonably consistent with each other. However, both sets of experimental results show smaller values of A_s/A_d than predicted by equations (9.2) and (9.3). This is particularly the case for small values of x, where the area of the doublet is much greater than that predicted for a random distribution of Fe impurity atoms. This feature indicates that the Fe impurity atoms that form a supersaturated solution in fcc aluminum, do not occupy lattice sites at random but show a strong tendency for clustering. These Mössbauer results illustrate the information that can be gained from this technique and illustrate that such results can complement those obtained from other methods such as x-ray diffraction and provide a more comprehensive view of the microstructure of these alloys.

9.6.2 FeM alloys

Figure 9.18 summarizes isomer shift studies of dilute impurities in an Fe host of the form FeM. These data show the same general trends as the Fe-impurity data illustrated in figure 9.14. This suggests that the same charge transfer mechanism that described this previous data is suitable for the description of the data in figure 9.18. It is interesting to note, however, that the range of values across the 3d series shown in figure 9.14 is significantly greater than the range of values shown in figure 9.18. In a random dilute (bcc) MFe alloy (e.g., $M_{0.98}Fe_{0.02}$) about 85% of ^{57}Fe probe atoms have eight nearest neighbor host atoms and the remainder mostly have seven host and one Fe neighbors (see equation (9.2). In a dilute FeM alloy (e.g., $Fe_{0.98}M_{0.02}$) most (~85%) ^{57}Fe probe atoms would have eight Fe neighbors and most of the rest would have seven Fe neighbors and one impurity neighbor. Thus, the changes to the Fe electronic structure as a function of impurity electronic structure would be much less for the FeM alloy than for the MFe alloy.

Since an Fe host alloy with dilute impurities would be ferromagnetic, the Mössbauer effect is a sensitive method for looking at the effects of impurities on the magnetic properties of iron. Vincze and Campbell (1973) have reported Mössbauer effect studies of Fe with a variety of dilute impurities, i.e., *Fe*M. Typical room temperature ^{57}Fe Mössbauer effect spectra from this study are shown in figure 9.19. Spectra showed one or two resolved satellite peaks on the outer

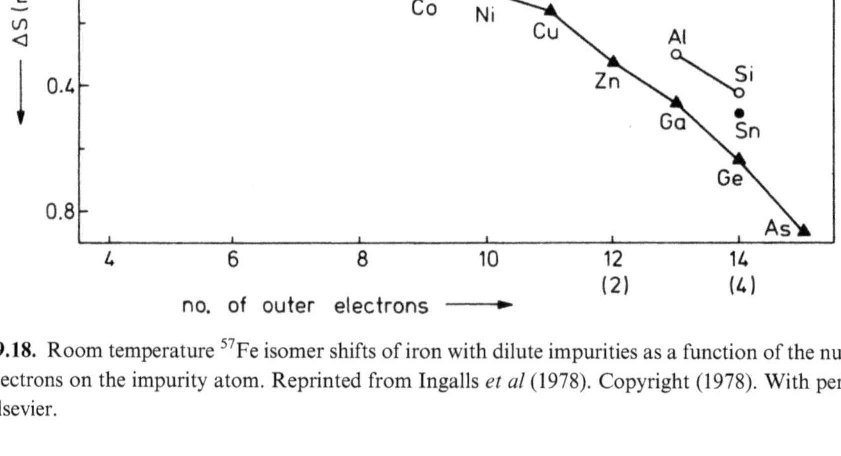

Figure 9.18. Room temperature ^{57}Fe isomer shifts of iron with dilute impurities as a function of the number of outer electrons on the impurity atom. Reprinted from Ingalls *et al* (1978). Copyright (1978). With permission from Elsevier.

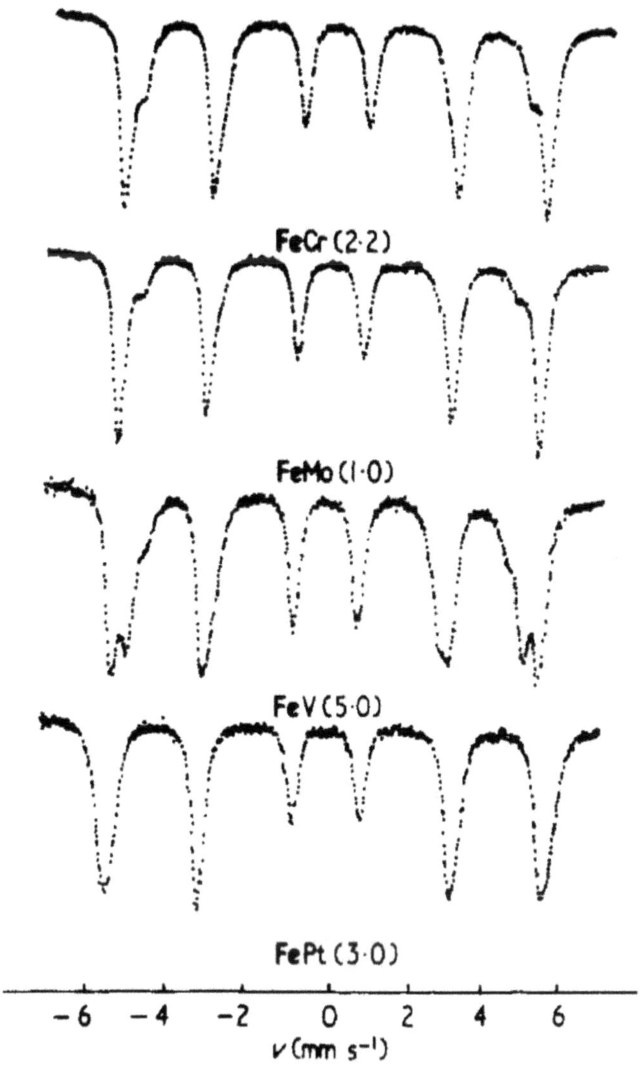

Figure 9.19. Room temperature ^{57}Fe Mössbauer effect spectra of some Fe alloys with dilute transition metal impurities. The percentage of impurity atoms is shown in (). From Vincze and Campbell (1973), copyright IOP Publishing. Reproduced with permission. All rights reserved.

absorption peaks of the Fe spectrum (e.g., for FeCr and FeV, respectively) or asymmetric broadening of the outer peaks (e.g., for FePt).

It is assumed that the different components of the ^{57}Fe Mössbauer spectra correspond to Fe probe atoms with different neighboring configurations of impurities. For an Fe alloy containing a concentration of x randomly distributed impurities, the probability of having an Fe atom with n impurities in the first nearest neighbor shell and m impurities in the kth nearest neighbor shell, is given by the binomial distribution as

$$P(c_1, c_k, n, m, x) = \binom{c_1}{n} x^n (1 - x)^{c_1 - n} \binom{c_k}{m} x^m (1 - x)^{c_k - m}$$

where c_1 and c_k are the coordination numbers of the first and kth shells, respectively. For Fe with the bcc structure $c_1 = 8$, $c_2 = 6$, $c_3 = 12$, etc, Vincze and Campbell (1973) found that effects from the first three nearest neighbor shells were observable (at least for some impurities) in the measured Mössbauer spectra.

In order to determine the overall effect of impurities on the hyperfine magnetic field in the FeM alloy, it was assumed that the effects of impurities were cumulative. That is, the hyperfine field at an ^{57}Fe probe site which has n first nearest neighbor impurity atoms and m kth nearest neighbor impurity atoms can be expressed as

$$H(n, m) = H_0 + n\Delta H_1 + m\Delta H_k$$

where H_0 is the component of the Fe hyperfine field that is independent of impurity concentration and ΔH_1 and ΔH_k are the change in hyperfine field per impurity atom in the first and kth nearest neighbor shells, respectively.

Analysis of Mössbauer spectra allow for the determination of ΔH_1 and ΔH_2 and, when resolvable ΔH_3, as a function of the impurity element. This analysis allows for the determination of the average hyperfine field and its dependence on the impurity concentration. Changes in the average hyperfine field as a result of impurities in Fe are the result of two contributions, core polarization arising directly from changes of the d-electron moment on Fe and conduction electron polarization. Vincze and Campbell (1973) have calculated the change in average Fe hyperfine field per unit impurity concentration, $\left[\frac{d\overline{H}}{dx}\right]_{\text{calc}}$, as

$$\left[\frac{d\overline{H}}{dx}\right]_{\text{calc}} = A\left(\frac{d\overline{\mu}}{dx} + \mu_{\text{Fe}} + \mu_{\text{i}}\right) + B(\mu_{\text{i}} - \mu_{\text{Fe}}) \tag{9.4}$$

where $\overline{\mu}$ is the average atomic magnetic moment and μ_{i} and μ_{Fe} are the moments on the impurity and iron atoms, respectively. The coefficient A was determined from the behavior of pure Fe and the coefficient B was determined from the behavior of dilute Cu in Fe where it has been determined that $\mu_{\text{Cu}} = 0$. Impurity moments have been determined by neutron diffraction studies and have been reported in the literature as given in Vincze and Campbell (1973). The first term on the right-hand side of equation (9.4) is the result of induced d moment changes on the Fe sites and includes contributions from both core polarization and conduction electron polarization. The second term on the right-hand side of the equation gives the effect on the conduction electrons at the Fe site as a result of the impurity site moment.

Figure 9.20 shows the average hyperfine field changes for 3d impurities as determined from measured values from Mössbauer studies as described above and calculated values from equation (9.4). The figure shows a fairly linear relationship between the experimental and calculated values of the hyperfine field change per impurity atom and supports the approach to the description of the effects of impurities on Fe hyperfine field systematics as described above.

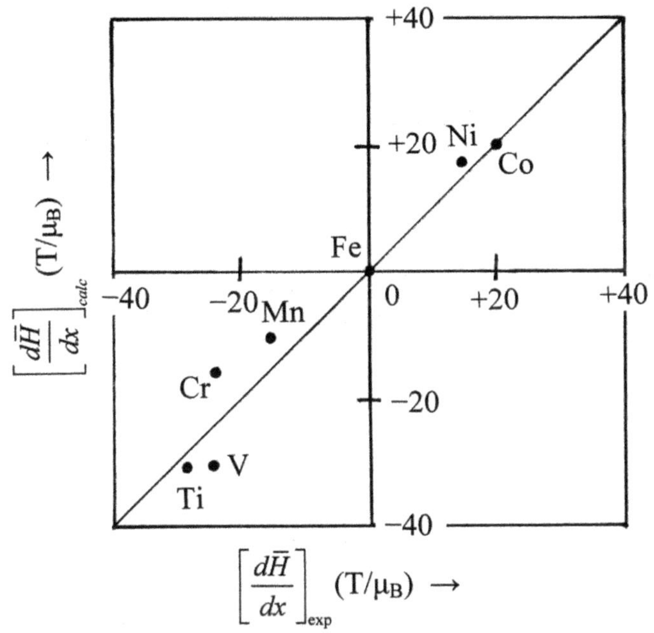

Figure 9.20. Measured (exp) and calculated (calc) average hyperfine field changes per unit impurity atom in an Fe host with dilute 3d impurity atoms. Plotted from data tabulated in Vincze and Campbell (1973).

9.7 Surface studies

As discussed in section 8.9, conversion electron Mössbauer spectroscopy is sensitive to the surface region of a sample. A simple approach to understanding the depth dependence of features of a material is through a comparison of the conversion electron backscatter Mössbauer spectrum and the x-ray backscatter Mössbauer spectrum. For example, for ^{57}Fe, the 7.3 keV conversion electrons have a typical penetration depth of about 100 nm while the 6.4 keV backscattered x-rays have a typical penetration depth of about 10 μm.

Sedláčková *et al* (2022) have reported an investigation of the effects of annealing on the surface features of some stainless steel alloys. CL20ES stainless steel powder was laser melted into samples of 25 × 25 × 3 mm^3. Major components of the alloy were Cr (16.5%–18.5%), Ni (10%–13%), Mo ((2.0%–2.5%), Mn (0%–2.0%) and Si (0%–1.0%), with the balance Fe. Mössbauer spectra were obtained with backscatter 6.4 keV x-rays and 7.3 keV conversion electrons.

Figure 9.21 shows room temperature x-ray backscatter spectra of samples annealed for different periods of time up to 128 h in air at 550 °C. The spectra all show the presence of a broadened singlet. Computer fits showed an isomer shift of about −0.10 mm s^{-1} relative to room temperature α-Fe characteristic of fcc-Fe. The broadening results from the presence of neighboring alloying atoms and can be fit with a quadrupole splitting of about 0.16 mm s^{-1}. Spectra also showed a fairly

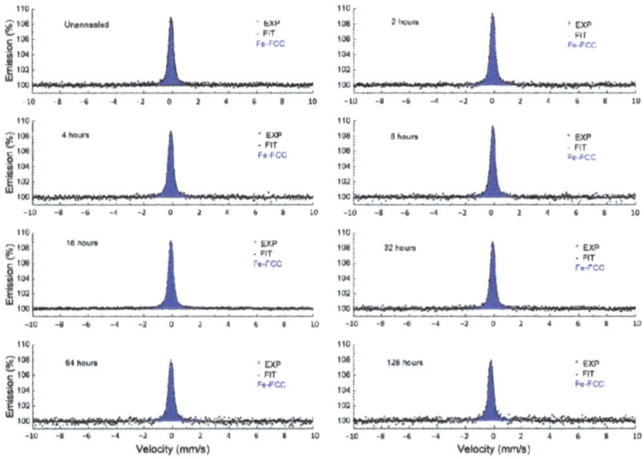

Figure 9.21. Room temperature x-ray backscatter ^{57}Fe Mössbauer spectra of samples annealed for different periods of time. From Sedláčková *et al* (2022) Copyright 2022 by the authors. CC BY 4.0.

systematic decrease in the amplitude of the emission peak as a function of increasing annealing time, as illustrated in figure 9.21.

Figure 9.22 shows room temperature conversion electron Mössbauer spectra of the same annealed samples. The spectra of all samples show a broadened singlet from fcc-Fe with Mössbauer parameters corresponding to those obtained from the x-ray backscatter spectra. Spectra of samples annealed for 2 h or longer showed an additional doublet component corresponding to Fe^{3+} with an isomer shift of around +0.36 mm s^{-1} and a quadrupole splitting of around 0.52 mm s^{-1}. Spectra of samples annealed for 4 h or longer also showed the presence of a sextet corresponding to bcc-Fe with an isomer shift near zero and a hyperfine field of 33.1 T to 33.9 T.

A comparison of figures 9.20 and 9.21 clearly shows that surface reactions occur on a distance scale of the order of the 100 nm penetration depth of the conversion electrons, but that these reactions do not measurably affect the sample on a depth scale comparable to the 10 μm penetration depth of the conversion x-rays. The decrease in the amplitude of the emitted x-rays with increasing annealing time, as shown in figure 9.21, suggests the formation of a non-Fe-containing surface layer that attenuates the backscattered x-rays. This is confirmed by the observation of a Cr_2O_3 surface layer using x-ray diffraction and energy dispersive x-ray spectroscopy.

The appearance of a Fe^{3+} doublet in the conversion electron spectra of annealed samples is consistent with the formation of a ferric oxide-containing phase near the surface. Bhattacharya *et al* (1997) have undertaken a Mössbauer effect study of mixed Fe_2O_3–Cr_2O_3 oxides and have reported quadrupole split doublets with isomer shifts between +0.33 mm s^{-1} and +0.36 mm s^{-1} and quadrupole splittings between 0.5 and 0.7 mm s^{-1}. The quadrupole split doublets seen in the spectra of annealed samples in figure 9.22 are consistent with the formation of surface $(Fe_{1-x}Cr_x)_2O_3$ during annealing.

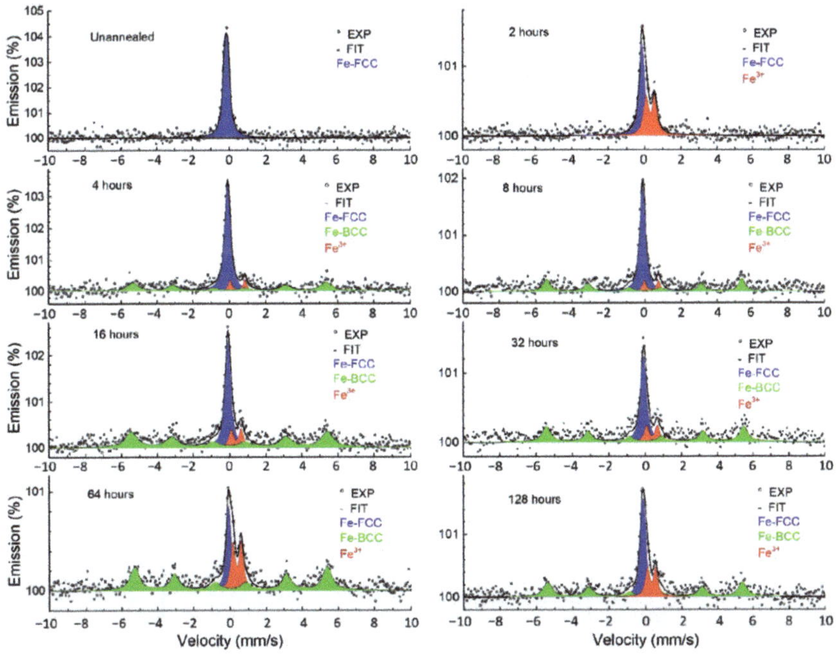

Figure 9.22. Room temperature ^{57}Fe conversion electron Mössbauer spectra of samples annealed for different periods of time. From Sedláčková *et al* (2022) Copyright 2022 by the authors. CC BY 4.0

Finally, the appearance of a Zeeman split sextet with hyperfine parameters consistent with α-Fe, indicates an fcc–bcc transformation in an iron-rich surface phase on a distance scale that is comparable to or less than the penetration depth of the conversion electrons. This is consistent with other similar studies (see e.g., Ivanova *et al* 2022).

The above example illustrates the ability of conversion electron Mössbauer spectroscopy to differentiate process in a surface region of a material and those that occur deeper in the sample. The ability of energy selective electron conversion Mössbauer spectroscopy to observe electrons from specific depths below the sample surface further increases the usefulness of this technique. An example of this approach is described below.

Shigematsu *et al* (1980) have prepared a layered Fe-containing sample by vacuum depositing a 25 nm thick iron film enriched to 95% in ^{57}Fe on a 91% ^{57}Fe enriched 310 stainless steel substrate. Depth selective conversion electron Mössbauer spectroscopy was conducted using an energy selective electron detector (see e.g., Bonchev *et al* 1969). Spectra as a function of conversion electron energy are shown in figure 9.23.

Since electrons which are liberated in an internal conversion process lose energy as they propagate through the sample, electrons which are produced deeper within the material have less energy when they emerge from the surface. Thus, measuring electrons with a specific energy less than the internal conversion energy, allows

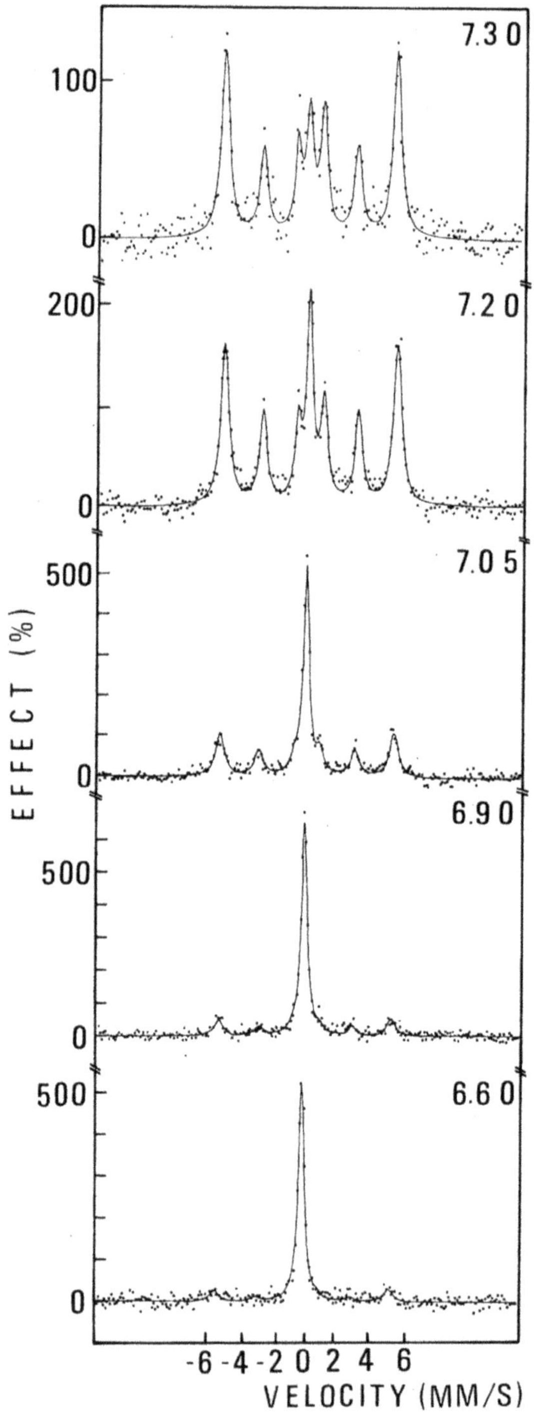

Figure 9.23. Energy selective conversion electron Mössbauer spectra of a 25 nm α-Fe layer on a 310 stainless steel substrate as described in the text. Reprinted with permission from Shigematsu *et al* (1980). Copyright 1980 by the American Physical Society.

Mössbauer spectra to be obtained that are characteristic of the sample at a particular depth below the surface.

The spectrum observed for 7.3 keV electrons is clearly dominated by a Zeeman split sextet characteristic of the α-Fe surface layer. In addition, a weaker singlet characteristic of the stainless steel substrate is seen. The spectrum also shows a doublet corresponding to Fe^{3+}, that originates from a thin iron oxide layer that formed on the surface when the sample was exposed to air. The doublet shows an isomer shift of +0.31 mm s^{-1} and a quadrupole splitting of 0.87 mm s^{-1} and may be compared to similar features observed by Sedláčková *et al* (2022) as discussed above.

Decreasing conversion electron energy, as indicated in figure 9.23, corresponds to increasing depth below the surface. The Mössbauer spectra as a function of depth show an increasing proportion of a singlet resulting from the paramagnetic stainless steel substrate. The spectrum corresponding to 6.6 keV conversion electrons is almost entirely (i.e. ~90%) from depths greater than 25 nm.

The two studies described above clearly illustrate the potential for conversion electron Mössbauer spectroscopy to probe near surface effects in solids.

9.8 Studies of battery materials

Secondary batteries, commonly referred to as rechargeable batteries, are used to power a wide variety of devices. These devices range from cell phones to battery electric vehicles. In order to improve the operation of these devices, much research has been undertaken in recent years to improve rechargeable battery technology. Such work has sought to improve battery characteristics such as energy density, recharge time and cost. The development of new electrode materials has been an important component of advanced battery development. Mössbauer spectroscopy is an effective and useful technique for investigating the properties of electrode materials that contain a suitable Mössbauer nuclide. While the vast majority of investigations have utilized ^{57}Fe or ^{119}Sn Mössbaurer transitions, results have also been reported for ^{121}Sb (Ionica *et al* 2005), ^{99}Ru (Mortemard be Boisse *et al* 2016) and ^{61}Ni (Segi *et al* 2016). The vast majority of this work has dealt with materials related to lithium-ion batteries, although some work has also been reported that is related to sodium-ion batteries (see e.g., Thorne *et al* 2013, Zheng *et al* 2016). Lippens (2021) has provided a recent review of the applications of Mössbauer spectroscopy to the study of battery materials. Here we begin with a brief overview of the operation of a rechargeable Li-ion battery.

The Li-ion battery stores energy during charge by using a current from an external power source to transport lithium ions from a positive electrode (cathode) through an electrolyte to a negative electrode (anode). This energy is released during discharge when a current in the opposite direction flows through an external circuit and the lithium ions are transported from the anode back through the electrolyte to the cathode. This situation is illustrated in figure 9.24.

Commonly used anode and cathode materials in lithium-ion cells are carbon (in the form of graphite) and lithium-metal-oxide (such as $LiCoO_2$), respectively. Thus, when the cell is charged, lithium is transported from the $LiCoO_2$ electrode to the

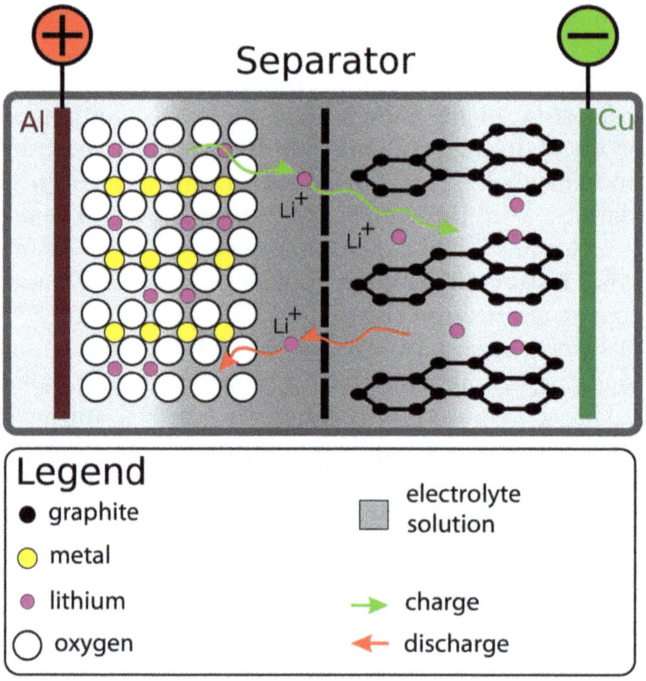

Figure 9.24. Diagram of a lithium-ion battery showing the movement of the Li-ions during charge and discharge. This [Schematischer Aufbau einer NMC-Lithium-Ionen-Zelle] image has been obtained by the author from the Wikimedia website where it was made available by Cepheiden (2020) under a CC BY-SA 2.0 licence. It is included within this chapter on that basis. It is attributed to Cepheiden [legend translated into English].

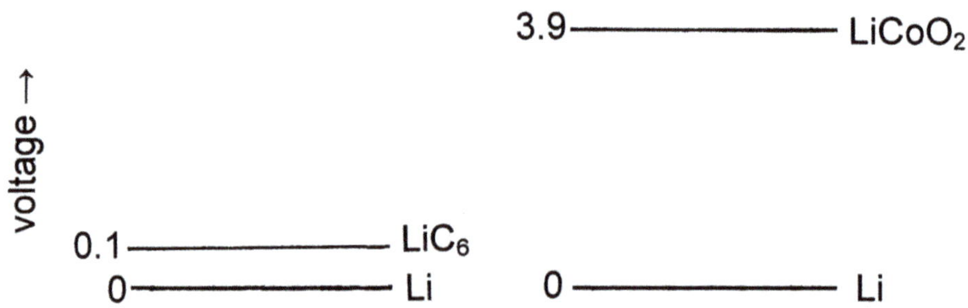

Figure 9.25. Chemical potential of LiC_6 (left) and $LiCoO_2$ (right) relative to elemental lithium metal. From Dunlap (2023) copyright IOP Publishing. Reproduced with permission. All rights reserved.

carbon electrode where it forms various lithium–carbon phases up to the maximum lithium capacity of LiC_6. The chemical potential of $LiCoO_2$ and LiC_6 are shown relative to elemental lithium in figure 9.25. Both compounds are at more positive potentials than elemental lithium and the cell voltage is determined by the potential difference between $LiCoO_2$ and LiC_6.

The details of the reactions that take place in the cell can be understood by looking at the processes that occur at each electrode (that is for each half cell) during the charge and discharge processes. We can write these as

$$LiCoO_2 \leftrightarrow CoO_2 + Li^{+} + e^{-} \text{ (cathode)} \tag{9.5}$$

and

$$C_6 + Li^{+} + e^{-} \leftrightarrow LiC_6 \text{ (anode).} \tag{9.6}$$

where the lithium ions travel through the electrolyte and the electrons travel through the external circuit. These processes are reversible and proceed from left to right during charge and from right to left during discharge. Equations (9.5) and (9.6) can be combined to yield the reaction for the complete electrochemical cell during charge and discharge as

$$LiCoO_2 + C_6 \leftrightarrow CoO_2 + LiC_6$$

Materials may be investigated as possible battery electrode materials by studying cells made from the material as one electrode and lithium metal as the other electrode. We can look at how this is done for a carbon anode. When a cell containing lithium versus carbon is constructed, it is initially in the charged state and is discharged as lithium is transported from the lithium electrode to the carbon electrode to form various lithium–carbon compounds. Thus, as illustrated in figure 9.26, the cell voltage can be measured as a function of the lithium content

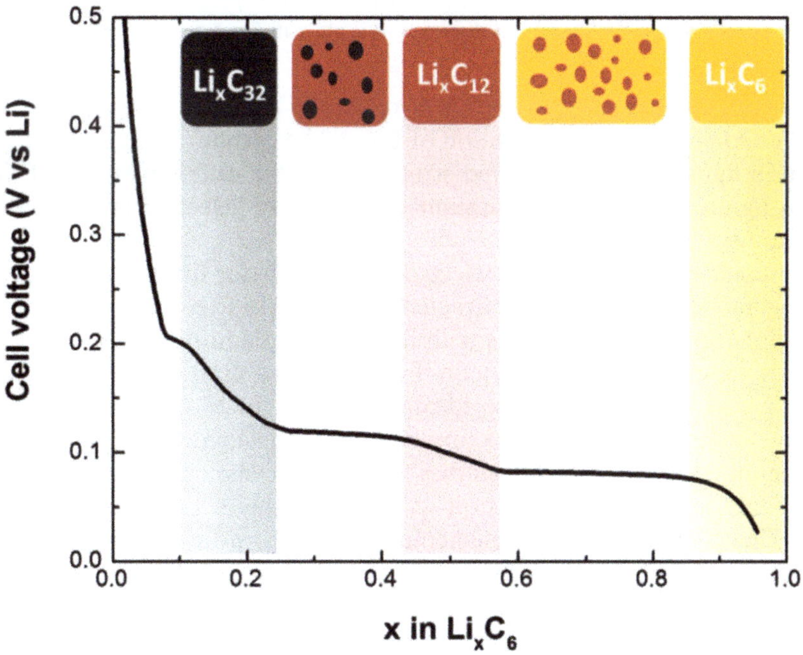

Figure 9.26. Voltage curve for the initial discharge of a lithium versus carbon cell showing the different stages. From Gallagher *et al* (2012) copyright IOP Publishing. Reproduced with permission. All rights reserved.

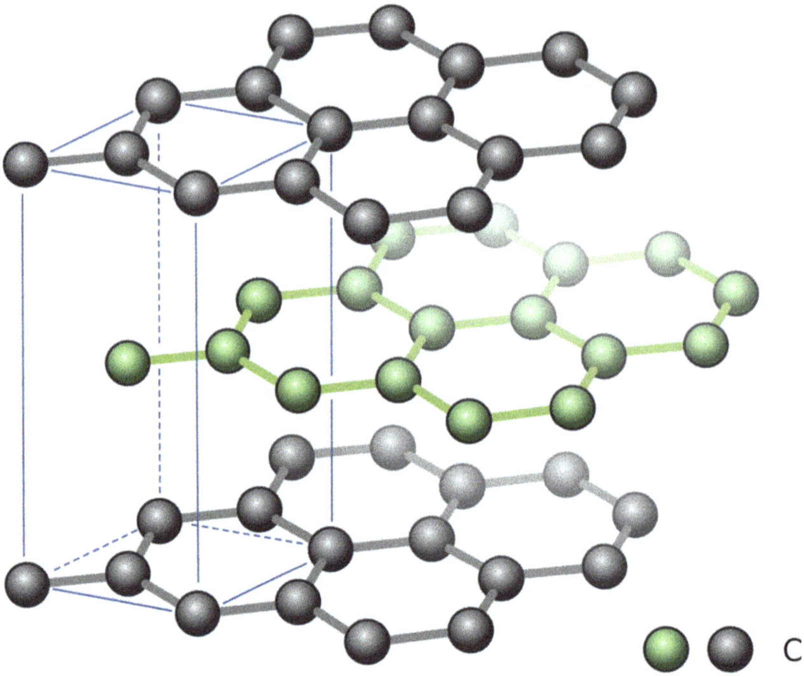

Figure 9.27. Layered structure of graphite showing hexagonal bonding within each layer. This [Balls and sticks model of AB stacked graphite with highlighted the unit cell composed of four atoms] image has been obtained by the author from the Wikimedia website where it was made available by Sikek (2015) under a CC BY-SA 4.0 licence. It is included within this chapter on that basis. It is attributed to Jozef Sivek [right panel of figure].

of the carbon electrode and the cell will reach the fully discharged state when the carbon electron reaches a composition of LiC_6. The carbon, in the form of graphite, as shown in figure 9.27, is a layered structure and the different stages of discharge shown in figure 9.26, represent lithium entering into selective layers between the carbon atoms.

In the search for high capacity electrode materials for lithium-ion batteries, tin-containing anode materials have attracted considerable attention. This is because tin can form intermetallic compounds with lithium up to a lithium-rich composition of $Li_{4.4}Sn$ (also referred to as $Li_{22}Sn_5$). Sn–Fe alloys have shown promise in this respect (e.g., Mao *et al* 1998) and are also ideal candidates for Mössbauer studies, as they contain two active nuclides. An example of such investigations is described below.

In general, Mössbauer effect studies of electrode materials can provide useful chemical and structural information at different stages of the charge–discharge process. Therse studies are complementary to x-ray diffraction studies and often more informative because the materials involved are frequently nanostructured. The simplest approach to such investigations is to cycle the cell to a particular point in the charge or discharge and to remove the electrode from the cell and collect a Mössbauer spectrum. This has the disadvantage that it is a tedious process if multiple points during the cycle are of interest. More importantly, however, it is

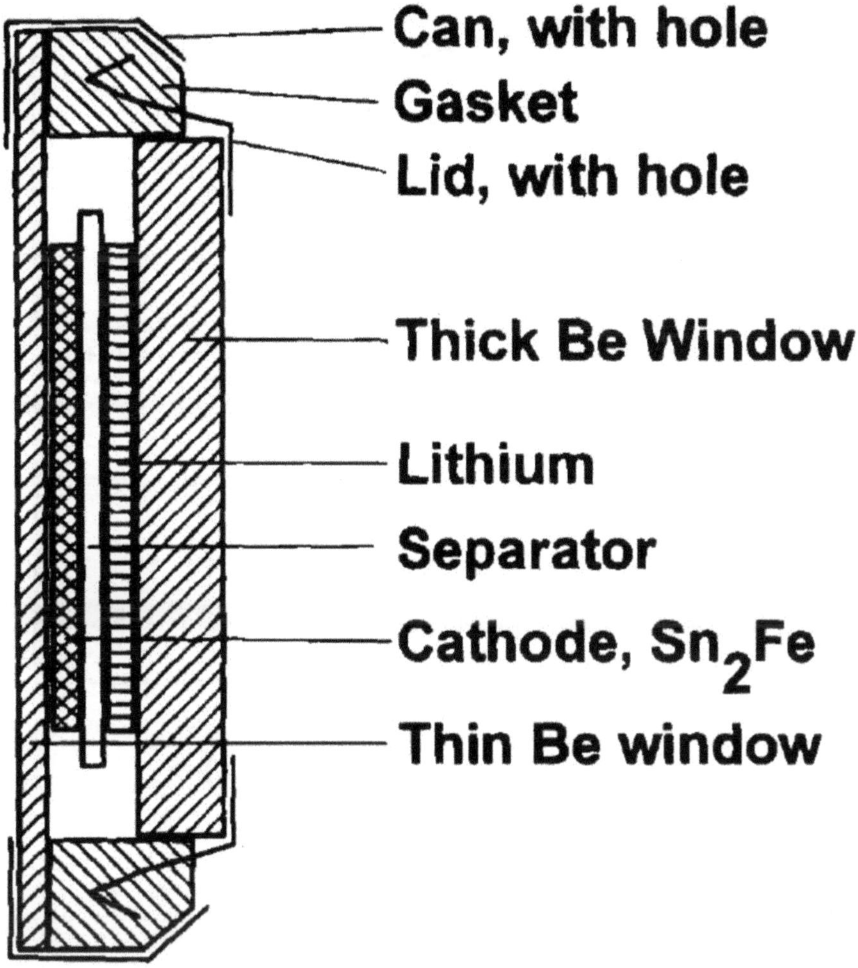

Figure 9.28. Diagram of an *in situ* Mössbauer effect cell based on a 2325 coin cell. From Mao *et al* (1998) copyright IOP Publishing. Reproduced with permission. All rights reserved.

possible that continuing chemical reactions can exist, even after the electrode material has been extracted from the cell and can affect the subsequent Mössbauer measurements. *In situ* Mössbauer effect studies, where spectra of the electrode material are obtained with the material in the cell during the cycling process, are not subject to these concerns. Several designs of *in situ* Mössbauer cells have been described in the literature (see references in Lippens 2021). A common design as described by Mao *et al* (1998) and Dunlap *et al* (1999a) utilizes a modified 2325 coin cell, as shown in figure 9.28. Circular holes are cut in the cell can and replaced with beryllium windows to allow the Mössbauer γ-rays to pass through the cell. These beryllium windows also act as the current collectors and the powdered cathode material is applied to the thin beryllium window and a lithium foil acts as the anode, as illustrated in the figure. In the as-prepared condition the cell is in the

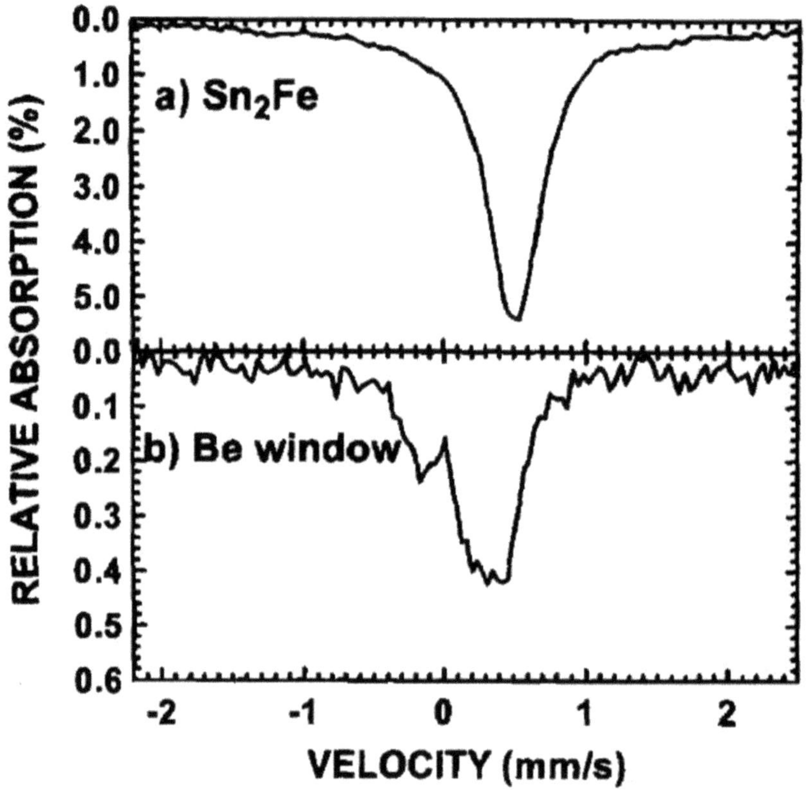

Figure 9.29. Room temperature ^{57}Fe Mössbauer effect spectra of (a) mechanically alloyed Sn_2Fe powder and (b) the thick Be window from the *in situ* cell. Note differences in vertical scales. From Mao *et al* (1998)

charged state and the cell is discharged as the lithium from the anode is transported to the cathode.

As an example of the use of the *in situ* cell, we consider a study of Sn_2Fe as reported by Mao *et al* (1998) and Dunlap *et al* (1999a). The ^{57}Fe Mössbauer spectrum of the Sn_2Fe powder, as prepared by mechanical alloying, is shown in figure 9.29 and exhibits a singlet. Although bulk Sn_2Fe is antiferromagnetic with a Néel temperature of 377 K and a room temperature Fe hyperfine field of 11.5 T, mechanically alloyed Sn_2Fe, as used in this study, has been shown to be superparamagnetic with a blocking temperature of about 20 K (see Dunlap *et al* 1999a and references therein). It is known that the beryllium windows contain a small fraction of Fe impurities that must be accounted for in the analysis of the *in situ* cell measurements. The ^{57}Fe Mössbauer spectrum of the beryllium window is shown in figure 9.29 and consists of a singlet due to isolated Fe impurities in the beryllium and a doublet corresponding to clusters of two or more Fe atoms.

^{57}Fe Mössbauer measurements were collected automatically with a duration of about 3 h each during cell cycling. The initial discharge required about 70–80 h

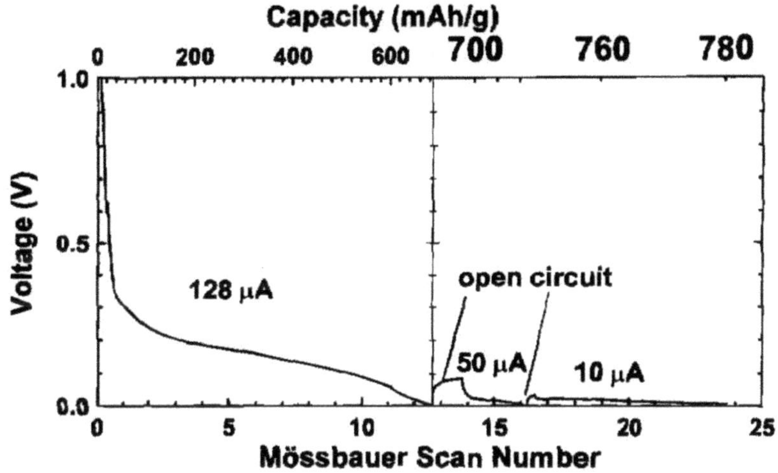

Figure 9.30. Cell voltage as a function of ^{57}Fe Mössbauer scan number (lower axis) and as a function of specific capacity (upper axis) during the initial discarge of a Li/Sn_2Fe cell. The constant current for each of the three segments of the discharge curve are indicated in the graph. From Mao *et al* (1998) copyright IOP Publishing. Reproduced with permission. All rights reserved.

yielding a total of 24 Mössbauer spectra over the duration of the discharge. The discharge curve as a function of Mössbauer spectrum number and cell capacity is illustrated in figure 9.30.

Figure 9.31 shows selected ^{57}Fe Mössbauer effect spectra obtained at various points during the initial discharge of the cell. Spectra exhibit the presence of three phases; Fe impurities in Be (which consists of a singlet and a doublet, as described above), Fe in Sn_2Fe (which consists of a singlet as shown in figure 9.29) and a doublet. It has been shown (Mao *et al* 1998, 1999) that the doublet corresponds to nanocrystalline Fe that is formed from Sn_2Fe that has reacted with lithium. The doublet nature of the Mössbauer spectrum of nanostructured Fe is consistent with previous work (Shull *et al* 1991). A quantitative assessment of phase development during cell cycling is possible from the analysis of Mössbauer spectra, as shown in figure 9.31. Because of the nanocrystalline Fe phase involved, this assessment is difficult from x-ray diffraction studies.

The capacity curve for the initial discharge of the Sn_2Fe versus Li cell associated with the ^{119}Sn Mössbauer measurements is shown in figure 9.32. Selected ^{119}Sn Mössbauer effect spectra obtained during this discharge are illustrated in figure 9.33. At the beginning of discharge the ^{119}Sn Mössbauer effect spectrum is a doublet with an isomer shift of 2.116 mm s^{-1} (relative to $CaSnO_3$) and a quadrupole splitting of 0.829 mm s^{-1}. This is consistent with expectations for superparamagnetic Sn_2Fe where the Sn atoms occupy anisotropic 8*h* sites in the tetragonal structure. During the initial discharge, the amplitude of the doublet decreases at the expense of a growing singlet due to Sn that has reacted with lithium. At the bottom of discharge, only the singlet component remains in the ^{119}Sn Mössbauer spectrum. Analysis of the spectrum in figure 9.33(d) gives an isomer shift of 1.80 mm s^{-1} relative to

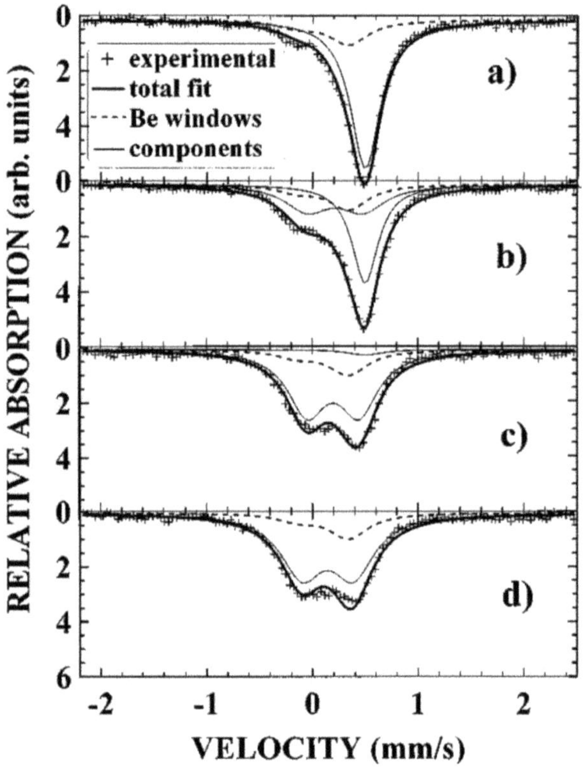

Figure 9.31. Selected ^{57}Fe Mössbauer effect spectra obtained taken during the initial discharge: (a) scan 1 (beginning of discharge), (b) scan 5, (c) scan 12 and (d) scan 24 (bottom of discharge). Scan numbers refer to those shown in figure 9.30. The fit to the Fe in the Be window is shown by the broken line and the singlet and doublet spectral components for the cell are shown by solid lines. Reprinted with permission from Dunlap *et al* (1999a). Copyright 1999 by the American Physical Society.

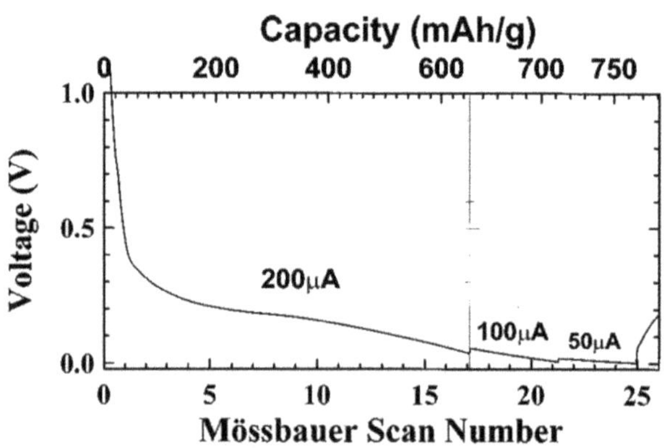

Figure 9.32. Cell voltage as a function of ^{119}Sn Mössbauer scan number (lower axis) and as a function of specific capacity (upper axis) during the initial discarge of a Li/Sn_2Fe cell. The constant current for each of the three segments of the discharge curve are indicated in the graph. Reprinted with permission from Dunlap *et al* (1999a). Copyright 1999 by the American Physical Society.

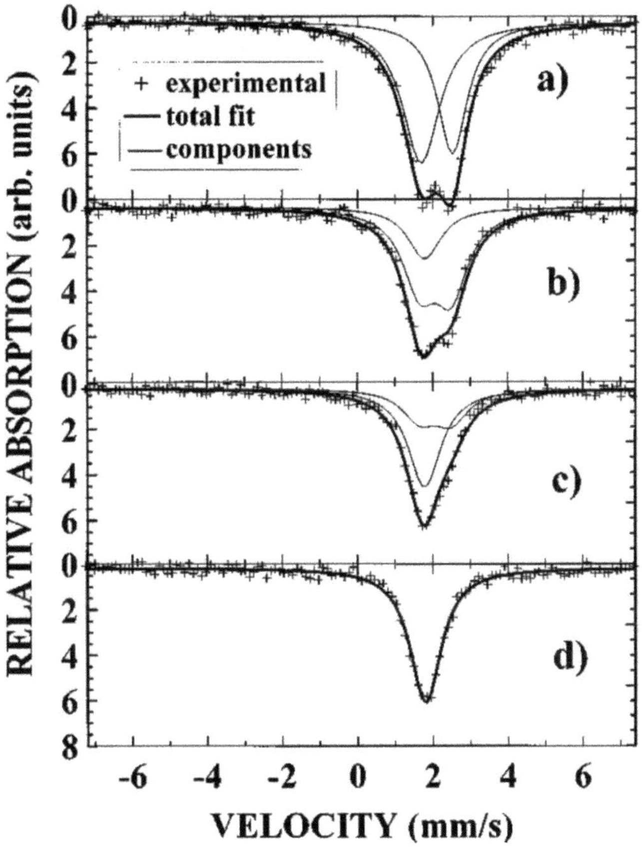

Figure 9.33. Selected ^{119}Sn Mössbauer effect spectra obtained taken during the initial discharge: (a) scan 1 (beginning of discharge), (b) scan 10, (c) scan 15 and (d) scan 25 (bottom of discharge). Scan numbers refer to those shown in figure 9.31. The fit to the spectral components for the cell are shown by solid lines. Reprinted with permission from Dunlap *et al* (1999a). Copyright 1999 by the American Physical Society.

$CaSnO_3$. On the basis of electrochemical considerations (e.g., Mao *et al* 1999), as well as x-ray studies of fully lithiated Sn_2Fe, we expect a composition of the Sn-containing phase at the bottom of discharge of around $Li_{4.4}Sn$. Dunlap *et al* (1999b) have provided ^{119}Sn Mössbauer effect spectra of Li–Sn reference samples that provide a means for interpreting *in situ* studies. These results give an isomer shift of the singlet $Li_{4.4}Sn$ spectrum of 1.837 mm s^{-1}, in reasonable agreement with the results from the *in situ* cell at the bottom of discharge.

Bibliography and suggestions for further reading

Benjah-bmm27 2007 *Graphite-Layers-Side-3D-Balls* (https://commons.wikimedia.org/wiki/File:Graphite-layers-side-3D-balls.png)

Bhattacharya A K, Hartridge A, Mallick K K, Majumdar C K, Das D and Chintalapudi S N 1997 An x-ray diffraction and Mössbauer study of nanocrystalline Fe_2O_3–Cr_2O_3 solid solutions *J. Mater. Sci.* **32** 557–60

Bonchev Z, Jordanov A and Minkova A 1969 Method of analysis of thin surface layers by the Mössbauer effect *Nucl. Instrum. Methods* **70** 36–40

Campbell I A and Blandin A 1975 Hyperfine fields at non-magnetic sites in metals *J. Magn. Magn. Mater.* **1** 1–10

Cepheiden 2020 *Schematischer Aufbau einer NMC-Lithium-Ionen-Zelle* https://commons.wikimedia.org/wiki/File:Li-Ion-Zelle_(NMC-Carbon,_Schema).svg

Coey J M D 1980 Clay minerals and their transformations studied with nuclear techniques: the contribution of Mössbauer spectroscopy *At. Energy Rev.* **18** 73–124

Coey J M D 1984 Mössbauer spectroscopy of silicate materials *Mössbauer Spectroscopy Applied to Inorganic Chemistry* ed G I Long (New York: Plenum) pp 443–509

Datta K J *et al* 2017 Synthesis of flower-like magnetite nanoassembly: application in the efficient reduction of nitroarenes *Sci. Rep.* **7** 11585

Dunlap R A 1982 Sn and Fe hyperfine field distributions in the amorphous ferromagnet $Fe_{80-x}Sn_xB_{20}$ *Solid State Commun.* **43** 57–60

Dunlap R A 1984 Hyperfine field distributions in some amorphous $(Fe,Ni,Cr)_{79}SnB_{20}$ alloys *J. Phys. F: Metal Phys.* **14** 549–53

Dunlap R A 2023 *Transportation Technologies for a Sustainable Future—Renewable Energy Options for Road, Rail, Marine and Air Transportation* (Bristol: IOP Publishing)

Dunlap R A, Lloyd D J, Christie I A, Stroink G and Stadnik Z M 1988 Physical properties of rapidly quenched Al–Fe alloys *J. Phys. F: Met. Phys.* **18** 1329–41

Dunlap R A, Mao O and Dahn J R 1999a Application of *in situ* Mössbauer effect methods for the study of electrochemical reactions in lithium ion battery electrode materials *Phys. Rev.* B **59** 3494–500

Dunlap R A, McGraw J D and Farrell S P 2006 A Mössbauer effect study of structural ordering in rapidly quenched Fe–Ga alloys *J. Magn. Magn. Mater.* **305** 315–20

Dunlap R A, Obrovac M N, Li J, Smith A, Hatchard T D, Sanderson R J and Dahn J R 2010 Application of *in situ* ^{119}Sn and ^{57}Fe Mössbauer effect spectroscopy to the study of lithium-ion battery electrode materials *Möss. Eff. Ref. Data J.* **33** 32–5

Dunlap R A, Small D A, MacKay G R, O'Brien J W, Dahn J R and Cheng Z H 2000 Materials preparation by ball milling *Can. J. Phys.* **78** 211–29

Dunlap R A, Small D A, MacNeil D D, Obrovac M N and Dahn J R 1999b A Mössbauer effect investigation of the Li–Sn system *J. Alloys Compd.* **289** 135–42

Gallagher K G, Dees D W, Jansen A N, Abraham D P and Kang S H 2012 A volume averaged approach to the numerical modeling of phase-transition intercalation electrodes presented for Li_xC_6 *J. Electrochem. Soc.* **159** A2029

Gaudet J M, Hatchard T D, Farrell S P and Dunlap R A 2008 Properties of Fe–Ga based powders prepared by mechanical alloying *J. Magn. Magn. Mater.* **320** 821–9

Greenwood N N and Gibb T C 1971 *Mössbauer Spectroscopy* (London: Chapman and Hall)

Ingalls R, Van der Woude F and Sawatzky G A 1978 Iron and nickel *Mössbauer Isomer Shifts* ed G K Shenoy and F E Wagner (Amsterdam: North-Holland) ch 7 pp 361–430

Ionica C M, Lippens P E, Olivier-Fourcade J and Jumas J C 2005 Study of Li insertion mechanisms in transition metal antimony compounds as negative electrodes for Li-ion battery *J. Power Sources* **146** 478–81

Ivanova T, Mashlan M, Ingr T, Doláková H, Sarychev D and Sedláčková A 2022 Mössbauer spectroscopy for additive manufacturing by selective laser melting *Metals* **12** 551

Kalvius M and Kienle P 2012 *The Rudolf Mössbauer Story: His Scientific Work and Its Impact on Science and History* (Cham: Springer)

Khong J C, Daisenberger D, Burca G, Kockelmann W, Tremsin A S and Mi J 2016 Design and characterisation of metallic glassy alloys of high neutron shielding capability *Sci. Rep.* **6** 36998

Le Caër G and Dubois J M 1979 Evaluation of hyperfine parameter distributions from overlapped Mössbauer spectra of amorphous alloys *J. Phys. E: Sci. Instrum.* **12** 1083–90

Lippens P E 2021 Application of Mössbauer spectroscopy to Li-ion and Na-ion batteries *Modern Mössbauer Spectroscopy—New Challenges Based on Cutting-Edge Techniques* ed Y Yoshida and G Langouche (Singapore: Springer Nature) ch 7 pp 319–79

Mao O, Dunlap R A, Courtney I A and Dahn J R 1998 In-situ Mössbauer effect studies of the electrochemical reaction of lithium with mechanically alloyed Sn_2Fe *J. Electrochem. Soc.* **145** 4195–202

Mao O, Dunlap R A and Dahn J R 1999 Mechanically alloyed Sn–Fe–(C) powders as anode materials for Li-ion batteries I: The Sn_2Fe–C system *J. Electrochem. Soc.* **146** 405–13

Miglierini M B 2021 Radiation effects in amorphous metallic alloys as revealed by Mössbauer spectrometry: part I. Neutron irradiation *Metals* **11** 845

Miglierini M B, Dekan J, Cesnek M, Janotová I, Švec P, Bujdoš M and Kohout J 2020 Hyperfine interactions in Fe/Co–B–Sn amorphous alloys by Mössbauer spectrometry *J. Magn. Magn. Mater.* **500** 166417

Miglierini M B and Rusakov V S 2010 ^{57}Fe and ^{119}Sn Mössbauer effect study of Fe–Sn–B amorphous alloys *AIP Conf. Proc.* **1258** 29–38

Mortemard de Boisse B *et al* 2016 Intermediate honeycomb ordering to trigger oxygen redox chemistry in layered battery electrode *Nat. Commun.* **7** 11397

Nasu S, Gonser U, Shingu P H and Marakami Y Y 1974 ^{57}Fe Mössbauer spectra in splat quenched Al-0.5, 1, 3 and 5 at% Fe alloys *J. Phys. F: Metal Phys.* **4** L24–8

Obrovac M N, Dunlap R A, Sanderson R J and Dahn J R 2001 The electrochemical displacement reaction of lithium with metal oxides *J. Electrochem. Soc.* **148** A576–88

Persoons R M, De Grave E and Vandenbergh R E 1990 Mössbauer study of Co-substituted magnetite *Hyperfine Interact.* **54** 655–60

Pound R V and Rebka Jr G A 1959 Gravitational red-shift in nuclear resonance *Phys. Rev. Lett.* **3** 439–41

Principi G 2020 The Mössbauer effect: a romantic scientific page *Metals* **10** 992

Prokaryotic Caspase Homolog 2023 *Pound and Rebka Experimental Arrangement* (https://commons.wikimedia.org/wiki/File:Pound_and_Rebka_experimental_arrangement.svg)

Reissner M 2021 Mössbauer spectroscopy in external magnetic fields *Modern Mössbauer Spectroscopy—New Challenges Based on Cutting Edge Techniques* ed Y Yoshida and G Langouce (Singapore: Springer Nature) pp 381–444

Rusanov V, Chakarova K, Winkler H and Trautwein A X 2009 Mössbauer and X-ray fluorescence measurements of authentic and counterfeited banknote pigments *Dyes Pigm.* **81** 254–8

Sedláčková A, Ivanova T, Mashlan M and Doláková H 2022 Phase changes in the surface layer of stainless steel annealed at a temperature of 550 °C *Materials* **15** 8871

Segi S, Masuda R, Kobayashi Y, Tsubota T, Yoda Y and Seto M 2016 Synchrotron radiation-based ^{61}Ni Mössbauer spectroscopic study of $Li(Ni_{1/3}Mn_{1/3}Co_{1/3})O_2$ cathode materials of lithium ion rechargeable battery *Hyperfine Interact.* **237** 1–5

Shigematsu T, Pfannes H D and Keune W 1980 Depth-selective conversion-electron Mössbauer spectroscopy *Phys. Rev. Lett.* **45** 1206–9

Shull R D, Ritter J J and Swartzendruber L J 1991 Change in magnetic state of Fe+silica gel nanocomposites due to low temperature treatment in ammonia *J. Appl. Phys.* **69** 5144–6

Sivek J 2015 *Balls and Sticks Model of AB Stacked Graphite with Highlighted the Unit Cell Composed of 4 Atoms* (https://commons.wikimedia.org/wiki/File:Balls_and_sticks_model_of_AB_stacked_graphite_with_other_layer_stackings.svg)

Thorne J S, Dunlap R A and Obrovac M N 2013 Structure and electrochemistry of $Na_xFe_xMn_{1-x}O_2$ ($1.0 \leqslant x \leqslant 0.5$) for Na-ion battery positive electrodes *J. Electrochem. Soc.* **160** A361–7

van der Woude F and Sawatzky G A 1974 Mössbauer effect in iron and dilute iron based alloys *Phys. Rep.* **12** 335–74

Vincze I and Campbell I A 1973 Mössbauer measurements in iron based alloys with transition metals *J. Phys. F: Metal Phys.* **3** 647–63

Wertheim G K, Guggenheim H J and Buchanan D N 1968 Sub lattice magnetization in FeF_3 near the critical point *Phys. Rev.* **169** 465–9

Yoshida Y and Langouche G (eds) 2013 *Mössbauer Spectroscopy: Tutorial Book* (Heidelberg: Springer)

Zheng L, Dunlap R A and Obrovac M N 2016 The electrochemical reaction mechanism of tin phosphide with sodium *J. Electrochem. Soc.* **163** A1188–91

IOP Publishing

The Mössbauer Effect (Second Edition)

Richard A Dunlap

Chapter 10

Applications of Mössbauer spectroscopy to other fields

10.1 Introduction

In addition to the applications discussed in the previous chapter, Mössbauer effect spectroscopy has been applied to various other fields of research. These fields include the study of naturally occurring materials. The prevalence of iron in our environment makes ^{57}Fe Mössbauer effect spectroscopy, in particular, a useful tool for many such applications and these studies include both materials of terrestrial origin and extraterrestrial origin. The Mössbauer effect has also been a very useful tool for the study of archaeological artifacts, both for the investigation of the provenance of materials as well as the processing methods used. It has also found applications for the study of biological materials. Finally, in the present section we describe a novel use of Mössbauer spectroscopy to determine the provenance of ink on banknotes.

10.2 Mineralogical studies

Following along the approach described in the previous chapter, the Mössbauer effect is a very useful technique for understanding the properties of mineral species that contain a Mössbauer active element. Although iron is a common component of a large fraction of minerals, some minerals may also contain other Mössbauer active elements, frequently, Sn or Sb, which can be investigated. Here we focus on some iron-containing materials.

As in the case of Fe_3O_4, it is observed that iron in minerals generally exists in 2+ or 3+ valence states and in sites with various local symmetries. The different valence states can be readily identified on the basis of their isomer shifts by referring to figure 4.6. The most common types of iron sites that are observed in minerals are octahedral Fe^{3+} sites and both octahedral and tetrahedral Fe^{2+} sites. In many cases, there is no long-range magnetic ordering of the iron magnetic moments (at least not at room temperature) and spectra show combinations of singlets, or

doi:10.1088/978-0-7503-6039-5ch10

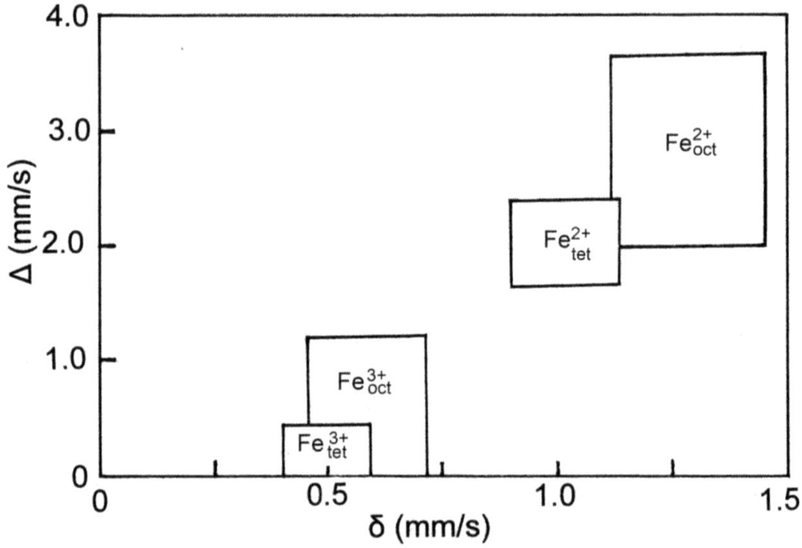

Figure 10.1. Typical ranges of the different sites that commonly occur in iron-containing silicates. For each range, the Fe valence and site symmetry. Data adapted from results given by Bancroft *et al* (1967).

(more commonly) quadrupole split doublets, that are representative of the valence states and symmetry of the iron environments. Figure 10.1 shows the typical range of isomer shifts and quadrupole splittings for high-spin Fe^{2+} and Fe^{3+} sites in naturally occurring silicates as a function of Fe valence state and coordination number. Sites are distinguished by their local coordination, four for tetrahedral sites and six for octahedral sites.

Figure 10.2 shows the Mössbauer spectra of some iron-containing silicates. The distinction between Fe^{2+} and Fe^{3+} sites is quite clear in the figure and provides information about the bonding preferences of the iron atoms in these different materials. The further classification of sites as M1, M2, etc, is based on local crystallographic symmetry, particularly the geometry of neighboring hydroxyl ions, which mainly influences the quadrupole splitting through changes in the electric field gradient at the iron site (see Coey 1984).

The above example shows that Mössbauer effect, particularly in conjunction with x-ray diffraction studies, are an effective method of understanding the structures of iron-containing minerals. As well, the Mössbauer effect offers the additional ability to provide information about the preferential site occupancy of iron in the structure. Thus, Mössbauer effect studies are useful for understanding the properties of specific minerals. The study of specific known minerals can also provide a database of Mössbauer effect parameters to allow the Mössbauer effect to be used as an identification tool for unknown materials. This approach is described further in the discussion of extraterrestrial materials below.

The vast majority of Mössbauer effect studies of mineralogical samples have utilized ^{57}Fe. This is largely because of the prevalence of iron in naturally occurring

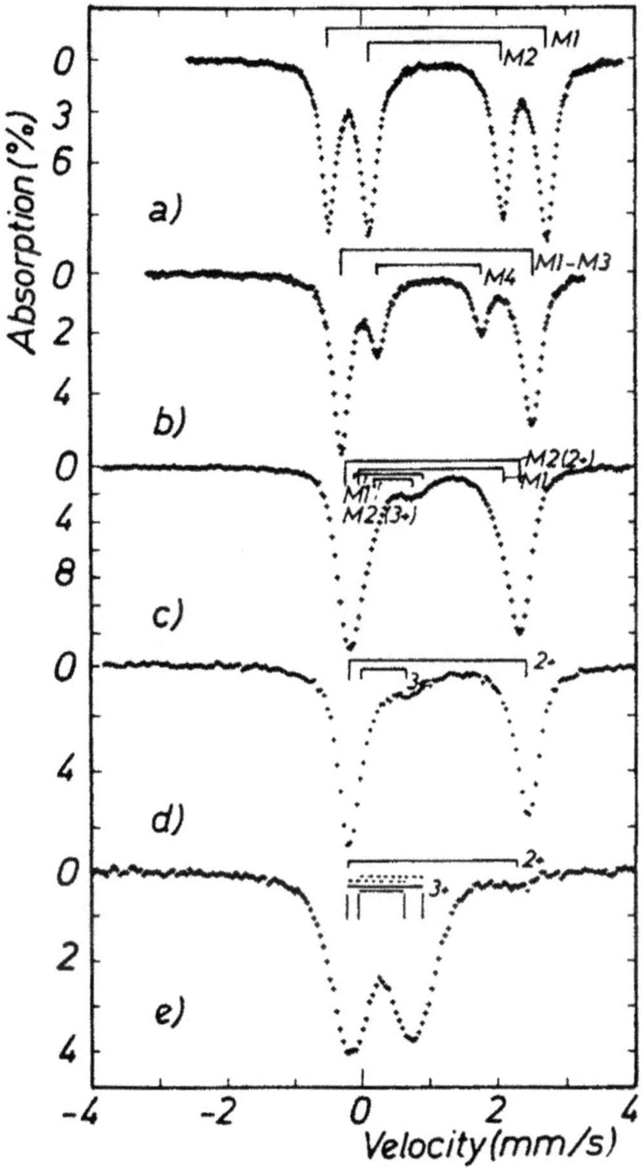

Figure 10.2. ^{57}Fe Mössbauer effect spectra of some iron-containing silicate minerals; (a) ferrous orthopyroxene [(Fe,Mg)SiO_3], (b) grunerite [$Fe_7Si_8O_{22}(OH)_2$], (c) biotite [K$(Fe,Mg)_3AlSi_3O_{10}(OH)_2$], (d) thuringite [$(Fe,Mg,Al)_6(Si,Al)_4O_{10}(OH)_8$] and (e) vermiculite [$(Fe,Mg)_3(Al,Si)_4O_{10}(OH)_2{\bullet}4H_2O$]. All spectra are at room temperature except (a) at 77 K. From Coey (1984) With permission of Springer.

materials. However, other Mössbauer effect nuclides have also provided useful information for understanding the properties of mineral samples. Figure 10.3 shows the result of a ^{119}Sn Mössbauer study of a naturally occurring tin-containing garnet. This spectrum shows an isomer shift of -0.14 mm s^{-1}, which indicates that the tin is

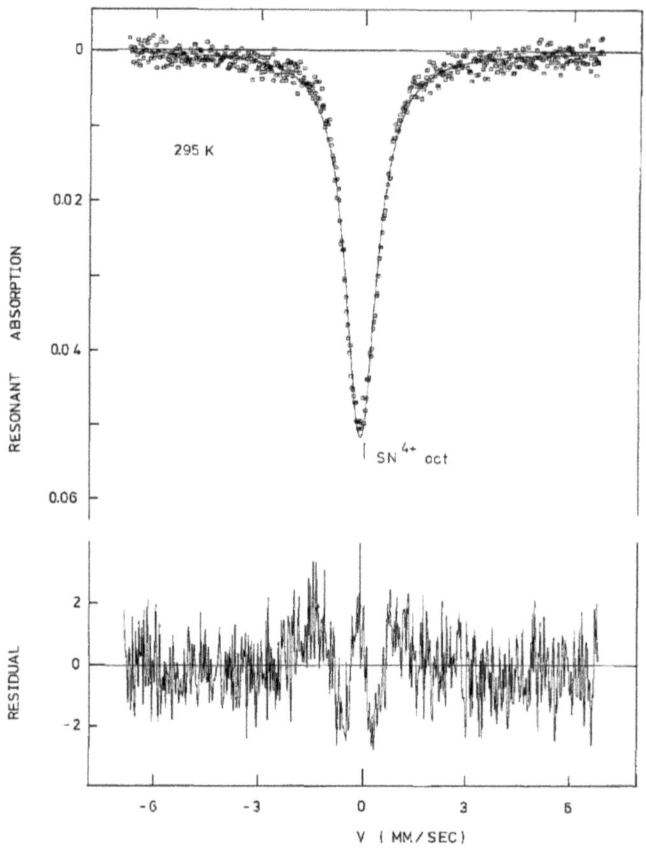

Figure 10.3. Room temperature ^{119}Sn Mössbauer effect spectrum of a green tin-andradite garnet sample from Davib Ost Mine, Karibib Constituency, Erongo Region, Namibia. From Amthauer *et al* (1979) With permission of Springer.

in a Sn^{4+} state. Based on site occupancy of Fe^{2+} and Fe^{3+} in similar materials, as obtained by ^{57}Fe Mössbauer measurements, it can be concluded that the Sn^{4+} resides in an octahedral site.

A fit of the spectrum in figure 10.3 to a single Lorentzian line gives a linewidth of 1.06 mm s^{-1}. This may be compared to the natural linewidth of the ^{119}Sn transition of 0.63 mm s^{-1}. The excess linewidth can largely be accounted for by fitting the spectrum to an unresolved quadrupole doublet. This fit gives a quadrupole splitting of $\Delta = 0.42$ mm s^{-1} and a more acceptable linewidth of 0.85 mm s^{-1}. This analysis indicates a slight asymmetry at the Sn sites caused by a distortion of the surrounding octahedral environment.

10.3 Studies of natural glasses

Naturally occurring glasses are silicate-based materials that can be formed by a variety of different methods. These glasses include volcanic glasses (obsidian),

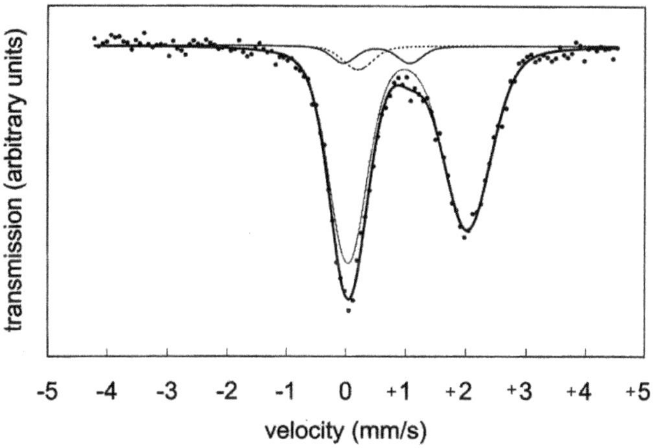

Figure 10.4. Room temperature ^{57}Fe Mössbauer effect spectrum of a Muong Nong tektite showing the total computer fit (heavy solid line) and the individual spectral components, Fe^{2+}_{oct} (thin solid line), Fe^{2+}_{tet} (medium solid line) and Fe^{3+} (broken line). Reprinted from Dunlap and Sibley (2004). Copyright (2004). With permission from Elsevier.

fulgurites, which are formed by lightning strikes and tektites and impact glasses, which are formed by meteorite impacts. Trinitite, which was formed during the testing of the first nuclear weapon test near Alamogordo, NM on 16 July 1945, is sometimes also included in this classification. All of these types of natural glasses contain some quantity of iron and have been the subject of ^{57}Fe Mössbauer effect investigations. In the present section we consider examples of Mössbauer effect spectra of tektites and fulgurites and describe the approach to analyzing these spectra and the information that can be extracted from such analyses. Obsidians have been studied by Mössbauer spectroscopy largely for the purpose of establishing provenance for materials used by ancient humans for tool manufacture. As such, these materials are discussed in section 10.5 on archeological studies.

A typical ^{57}Fe Mössbauer effect spectrum of a Muong Nong tektite from Eastern Laos is shown in figure 10.4. Muong Nong tektites, known as layered tektites because of their structure, are distinct from the more common splash-form tektites. The spectrum consists of a slightly asymmetric quadrupole split doublet. This spectrum is the result of contributions from Fe^{2+} and Fe^{3+} sites and the asymmetry is the result of correlations between the isomer shift, δ, and quadrupole splitting, Δ, not crystallographic orientation, as discussed in section 5.4. In the present case, the peaks have different depths but also have different widths so that the areas remain equal. Following the discussion in section 9.5, a simple analysis considers each site as a Gaussian distribution of quadruple splittings with a correlated isomer shift given by

$$\delta = \delta_0 + \alpha\Delta$$

where δ_0 and α are fitted correlation parameters. Rossano *et al* (1999) and Dunlap and Sibley (2004) have generalized this approach by considering the fitting of tektite

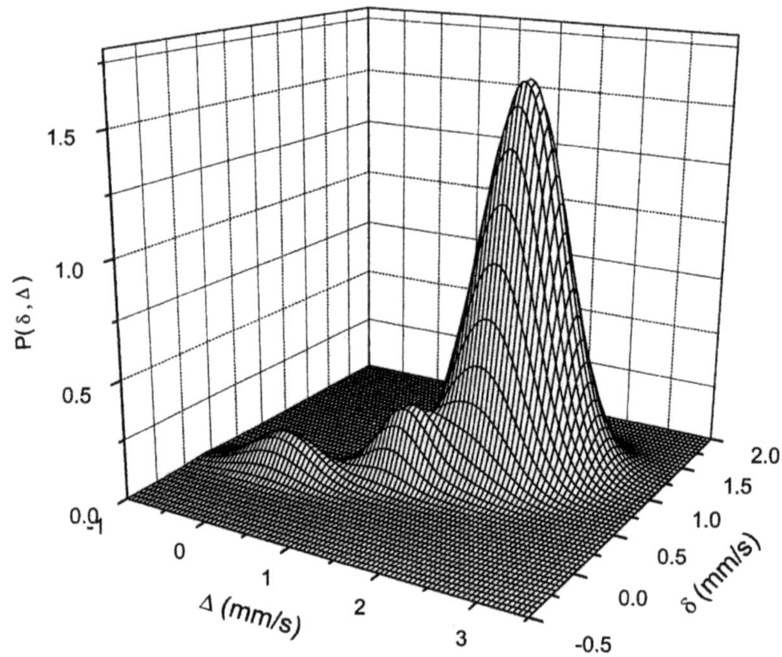

Figure 10.5. Three-dimensional representation of the total probability distribution P(δ,Δ) as a function of isomer shift, δ, and quadrupole splitting, Δ, for the spectrum shown in figure 10.3. Reprinted from Dunlap and Sibley (2004). Copyright (2004). With permission from Elsevier.

spectra using independent isomer shift and quadrupole splitting distributions. The results of such a fit for the spectrum in figure 10.4 are shown where three sites, Fe^{2+}_{oct}, Fe^{2+}_{tet} and Fe^{3+} have been utilized. Following along the lines of figure 10.1, the total distribution can be plotted as a function of isomer shift and quadrupole splitting, as shown in figure 10.5. This three-dimensional plot clearly shows the large peak in the probability distribution corresponding to the Fe^{2+}_{oct} sites. The Fe^{2+}_{tet} sites form a smaller peak for smaller isomer shift and quadrupole splitting and the Fe^{3+} peak is separated from the Fe^{2+} peaks.

An analysis of the parameters obtained from the fit to the tektite spectrum shown in figure 10.4, provides information about the properties of this tektite. Of particular importance, these results provide a measure of the Fe^{3+}/Fe^{2+} ratio. The Fe^{3+}/Fe^{2+} ratio in Mung Nong tektites is substantially larger than that in splash-form tektites. This difference provides evidence in favor of the common hypothesis that Muong Nong tektites solidified from the melt at a lower temperature than other types of tektites (Costa *et al* 2021).

Fulgurites are formed from terrestrial material, e.g., rock or sand, that has been struck by lightning. This process melts the surface matter, which is then rapidly cooled. ^{57}Fe Mössbauer effect studies of fulgurites and associated country rock (i.e., unmelted surface material from the same location) provides considerable information into the melting and solidification mechanism. Figure 10.6 shows Mössbauer

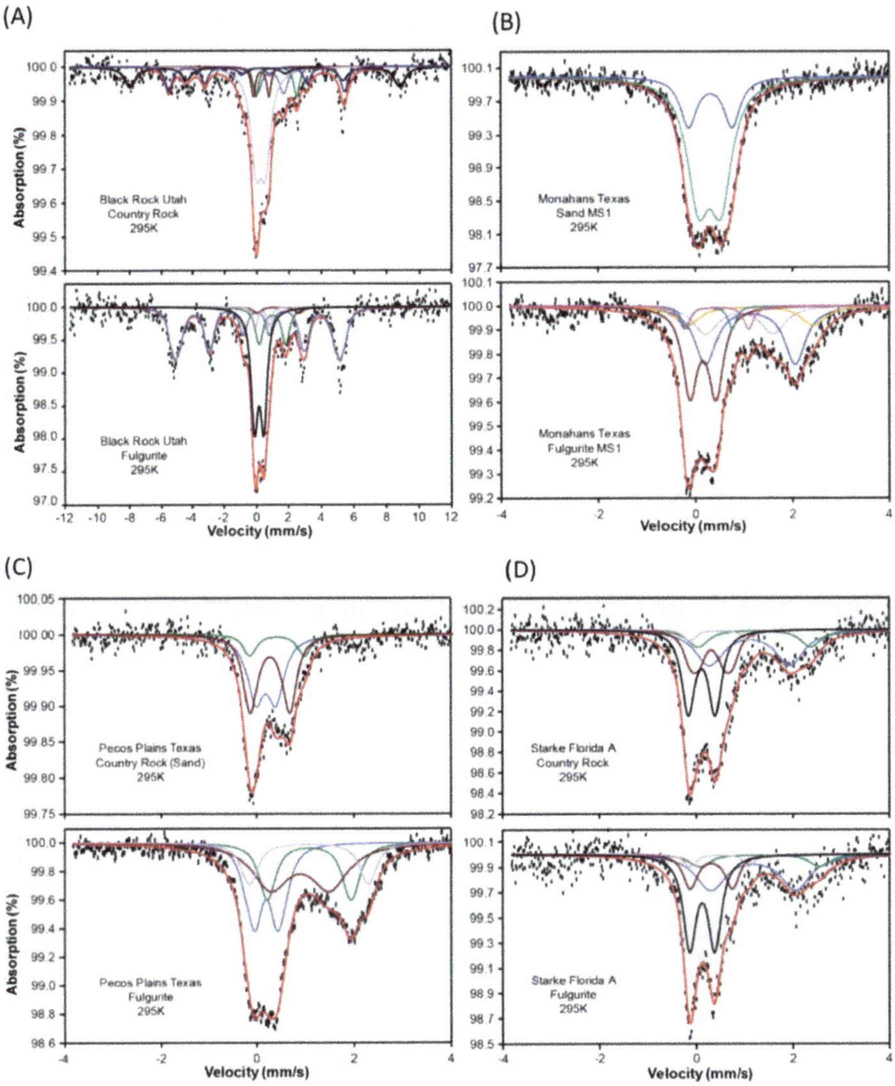

Figure 10.6. Room temperature ^{57}Fe Mössbauer spectra for some fulgurites and associated country rock specimens: (A) Black Rock, UT, (B) Monahans, TX, (C) Pecos Plains, TX and (D) Starke, FL. Reprinted from Roberts *et al* (2019). Copyright (2019). With permission from Elsevier.

studies of some fulgurite and associated country rock or sand samples, as reported by Roberts *et al* (2019). Spectra were fit to Fe^{2+}_{oct}, Fe^{2+}_{tet}, Fe^{3+}_{oct} and Fe^{3+}_{tet} quadruple split sites, as appropriate, along with magnetically spit components when observed. The Fe^{3+} features in the spectra can be readily observed by the components with small quadrupole splittings and small isomer shifts. The Fe^{2+} features can be seen as components with larger splittings and more positive isomer shifts, as expected from figure 10.1. In most of the cases illustrated in figure 10.6, as well as most other fulgurites reported in the same study, the target material shows a significant (average 66%) reduction of the Fe during the formation of the fulgurite. This is clearly evidenced by the relative increase in Fe^{2+} states in the fulgurite compared to the country rock sample. It is hypothesized that the reduction occurs as a result of a shockwave caused by the lightning strike that propagates through the material.

10.4 Investigations of extraterrestrial materials

The Mössbauer effect has been a useful technique for identifying and understanding the formation conditions of not only terrestrial minerals but materials of extraterrestrial origin. These investigations include a substantial number of studies of meteorites, as well as samples of lunar origin that were returned on moon missions in the 1960s and 1970s. The most recent approach to using Mössbauer effect spectroscopy to study extraterrestrial materials has been the incorporation of a functioning Mössbauer effect spectrometer on the NASA missions to Mars. In the present section, some representative results from these three types of studies are described.

10.4.1 Meteorite studies

Ordinary chondrites are the most common type of meteorite found on Earth. They are stony meteorites and are generally categorized by their total iron content as follows:

- *H chondrites* have the highest total iron content;
- *L chondrites* have lower total iron content;
- *LL chondrites* have the lowest total iron content.

They are further categorized by petrologic type (indicated by a number appended to the classification) which indicates the degree to which the material has been affected by the secondary processes of aqueous alteration or thermal metamorphism on the parent asteroid. Types 1 and 2 indicate increasing aqueous alteration while types 3–6 indicate increasing thermal metamorphosis.

^{57}Fe Mössbauer effect has been an effective means of identifying and quantifying iron phases in these types of meteorites. An example of some measurements on H and L chondrites is given below.

The identities of the chondrites that are discussed in this section are given in table 10.1. The samples include one H chondrite and three L chondrites. The room

Table 10.1. Locations, classifications and basic properties of meteorites described in this section.

Meteorite	Location	Classification	Total known mass (kg)	Year of fall
Beaver Creek	British Columbia, Canada	H4	14.1	1893
Shelburne	Ontario, Canada	L5	18.6	1904
Bruderheim	Alberta, Canada	L6	303	1960
Peace River	Alberta, Canada	L6	45.8	1963

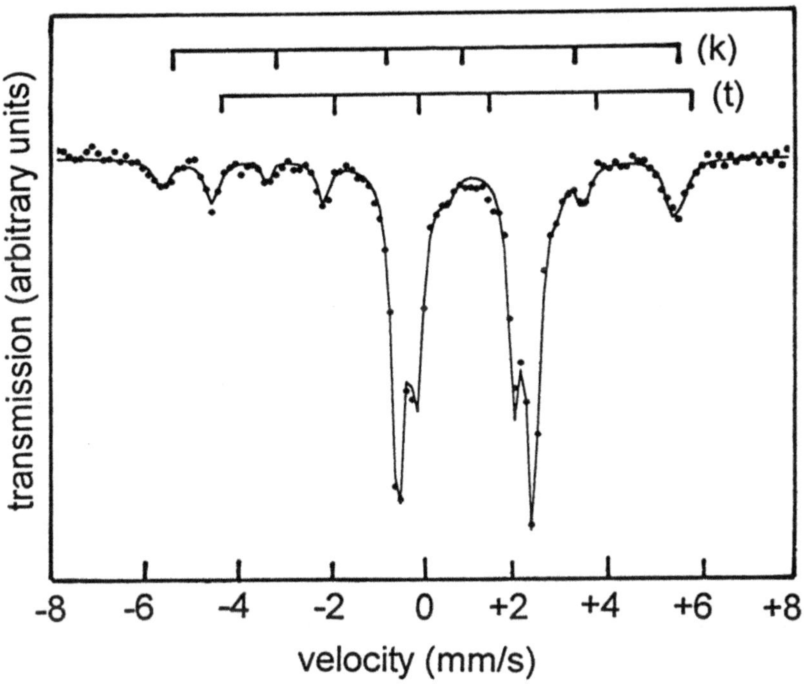

Figure 10.7. Room temperature ^{57}Fe Mössbauer effect spectrum of the L5 chondrite from Shelburne taken with a large velocity scale. The line positions for the ferromagnetic phases kamacite (k) and troilite (t) are shown.

temperature ^{57}Fe Mössbauer spectrum of one sample taken with a high range of transducer velocities is shown in figure 10.7. Spectra taken with this velocity range are suitable for viewing the spectral features associated with ferromagnetic Fe phases present in the sample. Figure 10.8 shows spectra of the four samples taken with a smaller velocity range. Such spectra are suitable for observing the details of the paramagnetic spectral components.

The Fe-containing phases that are commonly observed in ordinary chondrites are given in table 10.2. Identification of these phases is shown for the spectra in figures 10.6 and 10.7. Also shown in figure 10.7, is the location of a small Fe component due to Fe^{3+}. Fitted parameters are given in table 10.3.

The Fe hyperfine field in bcc $Fe_{100-x}Ni_x$ (kamacite) increases with Ni content for small values of x (Johnson *et al* 1961, 1963). The measured hyperfine fields for the four chondrites in table 10.3 are all slightly higher than that in pure α-Fe (33.0 T). These values are typical of kamacite with 4%–6% Ni. This is consistent with typical kamacite compositions found in chondrites.

The hyperfine field given in table 10.3 for the troilite component of the spectra is consistent with values reported in the literature (e.g., Barb *et al* 1982, Bonazzi *et al* 1992, Zhang *et al* 1994). The variations in the observed field may be the result of variations in the exact stoichiometry of the phase. There is some indication here, as

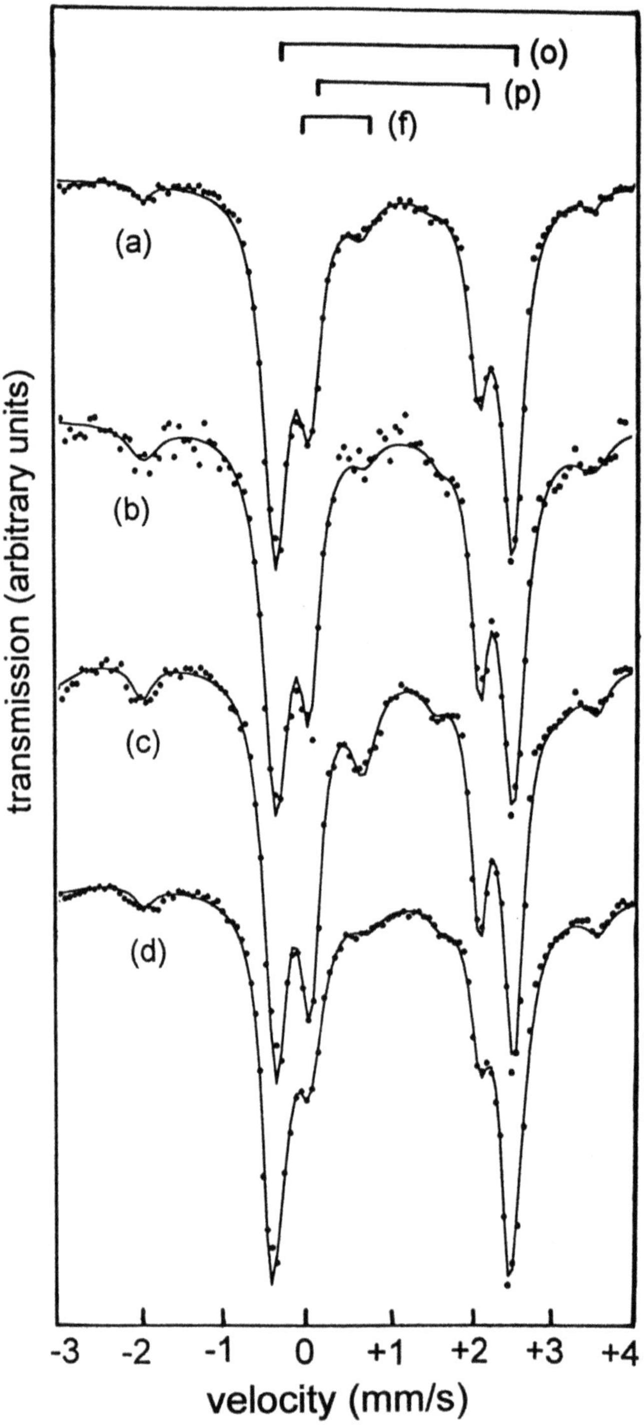

Figure 10.8. Room temperature ^{57}Fe Mössbauer effect spectrum of (a) Beaver Creek, (b) Shelburne, (c) Bruderheim, and (d) Peace River taken with a small velocity scale. The line positions for the paramagnetic phases olivine (o), pyroxene (p) and Fe^{3+} phases (f) are shown.

Table 10.2. Typical compositions of Fe-containing phases observed in ordinary chondrites.

Mineral	Typical composition
Kamacite	$Fe_{100-x}Ni_x$
Troilite	FeS
Pyroxene	$(Ca,Mg,Fe)_2Si_2O_6$
Olivine	$(Mg,Fe)_2SiO_4$

Table 10.3. Mössbauer effect parameters for Fe-containing phases identified in the room temperature ^{57}Fe spectra of several ordinary chondrites as presented in figures 10.7 and 10.8. Isomer shifts (δ) are given relative to room temperature α-Fe. Δ represents the quadrupole splitting for paramagnetic phases and the quadrupole shift (see section 7.5) for ferromagnetic phases. The sign of the quadrupole shift for Zeeman split spectra is determined by the direction of the shift of the inner absorption lines. A is the relative area of the spectral component. Uncertainties in velocities are ±0.005 mm s^{-1}, uncertainties in hyperfine fields are ±0.1 T and uncertainties in relative areas are ±2%.

Meteorite	Phase	δ (mm s^{-1})	Δ (mm s^{-1})	H (T)	A (%)
Beaver Creek	Kamacite	+0.021	−0.027	33.85	6
	Troilite	+0.675	+0.064	30.97	9
	Pyroxene	+1.152	2.133		23
	Olivine	+1.163	2.931		56
	Fe^{3+}	+0.402	0.695		6
Shelburne	Kamacite	+0.049	−0.038	33.93	11
	Troilite	+0.700	+0.066	31.27	14
	Pyroxene	+1.144	2.124		20
	Olivine	+1.171	2.938		50
	Fe^{3+}	+0.403	0.776		6
Bruderheim	Kamacite	+0.000	+0.041	33.40	13
	Troilite	+0.661	+0.092	31.14	15
	Pyroxene	+1.149	2.129		19
	Olivine	+1.162	2.957		42
	Fe^{3+}	+0.405	0.704		11
Peace River	Kamacite	+0.020	+0.010	33.80	7
	Troilite	+0.673	+0.066	31.36	17
	Pyroxene	+1.153	2.172	—	16
	Olivine	+1.163	2.951		56
	Fe^{3+}	+0.404	0.698		4

well as in the literature that the field in H chondrites may be slightly lower than in L chondrites. Troilite quadrupole shifts are positive and are typically greater in magnitude than those of kamacite. This is indicative of some degree of crystallographic distortion leading to a local symmetry that is lower than cubic.

The observed pyroxene isomer shift is around +1.15 mm s^{-1} for all samples and the observed quadrupole splitting is in the range 2.12–2.17 mm s^{-1}. Pyroxene forms a solid solution between diopside ($CaMgSi_2O_6$) and hedenbergite ($CaFeSi_2O_6$) with compositions $(CaMg_{1-x}Fe_x)Si_2O_6$. In general, the isomer shift of pyroxenes is relatively insensitive to composition and falls in the range of +1.11 to +1.17 mm s^{-1} (Coey 1984), consistent with the results in this section. The quadrupole splitting, however, varies with composition and the results in table 10.3 show a pyroxene composition in the diopside–hedenbergite phase diagram with very little Fe that falls close to diopside.

The isomer shifts for the olivine component of these samples are all around +1.16 mm s^{-1}. Similarly, the quadrupole splitting is around 2.94 mm s^{-1}. These values are consistent with literature values for olivines with low iron content. The olivines form a solid solution between forsterite (Mg_2SiO_4) and fayalite (Fe_2SiO_4) with compositions $(Mg_{1-x}Fe_x)_2SiO_4$. The results in table 10.3 show a composition with about $x = 0.2$–0.3.

The identification of the observed Fe^{3+} phase is not straightforward. Fe^{3+} can exist in a chondrite sample for one of two reasons, (1) it is an intrinsic component of the meteorite or (2) it formed after the meteorite fell by oxidation of a meteorite component, either Fe–Ni, FeS or a Fe^{2+} compound. The Mössbauer parameters of this phase reported in table 10.3, along with previously reported studies (e.g., Berry *et al* 1994, Zhang *et al* 1994), suggest that the Fe^{3+} observed in figure 10.8 was introduced during the grinding of the sample in preparation for Mössbauer measurements.

10.4.2 Studies of lunar samples

From 1969 to 1972 about 2200 samples with a total weight of 382 kg were returned from the moon by the Apollo missions (11, 12, 14, 15, 16 and 17). A number of these samples have been investigated by Mössbauer spectroscopy and a representative study is described below.

The lunar basalt sample 12018 was collected during the Apollo 12 lunar mission (14 November 1969–24 November 1969) in the Oceanus Procellarum region. Mössbauer studies of olivine that was separated from other components of the lunar basalt have been reported by Virgo and Hafner (1972). For comparison, a sample of fayalite, Fe_2SiO_4 (see above), from the type locality of Rockport, MA, was used. X-ray emission microanalysis showed the molecular fraction of fayalite in the lunar olivine to be 0.33 and in the Rockport sample to be 0.97. Figure 10.9 shows ^{57}Fe Mössbauer spectra of these two samples at around 270 °C. In order to understand the features that are shown in these spectra, it is useful to look in detail at the olivine structure. Figure 10.10 shows the olivine structure and, in particular, the locations of the two inequivalent octahedral sites for the Fe and Mg ions, designated M1 and M2 in the figure. At room temperature, the Mössbauer spectral components for Fe^{2+} at the M1 and M2 sites in olivines are not readily resolvable (see e.g., figure 10.8). However, the temperature dependence of the quadrupole splitting of the two sites is different, so that above about 200 °C, two doublets can be

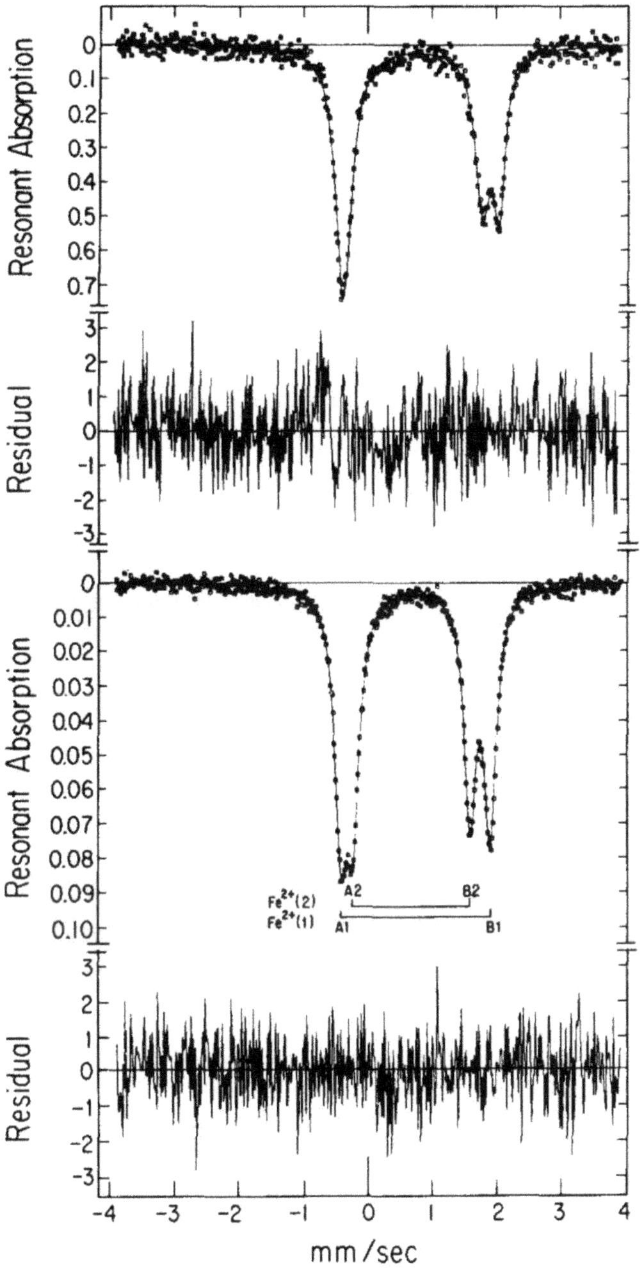

Figure 10.9. Room temperature ^{57}Fe Mössbauer spectra of (top) olivine from lunar basalt sample 12018 at 269 °C and (bottom) fayalite from Rockport, MA at 273 °C. Reprinted from Virgo and Hafner (1972). Copyright (1972). With permission from Elsevier.

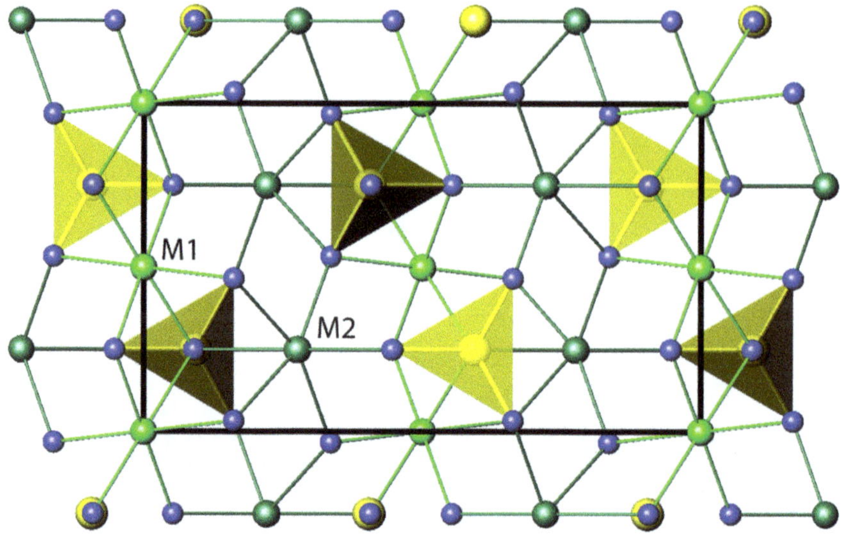

Figure 10.10. Olivine structure showing the locations of oxygen (blue), silicon (yellow) and iron, magnesium (green). The M1 and M2 sites for Fe, Mg are indicated and are shown by the lighter green atoms and darker green atoms, respectively. This [Crystal structure of Mg_2SiO_4 (olivine, Pnma) projected onto the (a, b) plane] image has been obtained by the author from the Wikimedia website https://commons.wikimedia.org/wiki/File:Olivine_Mg2SiO4.png (Perditax 2010), where it is stated to have been released into the public domain. It is included within this chapter on that basis.

Table 10.4. Mössbauer effect parameters (isomer shift relative to room temperature α-Fe, δ, and quadrupole splitting, Δ, for olivine from lunar basalt 12018 (at 269 °C) and Rockport fayalite (at 273 °C) from figure 10.9. x is the molecular fraction of fayalite as given in the text above. The area ratio, $B_1/(B_1 + B_2)$, as described in the text, is the average value over a range of temperatures (273 °C to 350 °C for 12018 and 269 °C to 420 °C for Rockport). The Fe^{2+} occupancy parameter, k, is defined in the text. Data are from Virgo *et al* (1972).

			M1		M2			
Sample	T (°C)	x	δ (mm s^{-1})	Δ (mm s^{-1})	δ (mm s^{-1})	Δ (mm s^{-1})	$B_1/(B_1+B_2)$	k
12018	269	0.33	+1.023	2.50	+0.902	2.26	0.411	0.58
Rockport	273	0.97	+1.005	2.548	+0.943	2.080	0.496	

resolved (Eibschütz and Ganiel 1967) and this is the case for the spectra in figure 10.9.

The fitted Mössbauer parameters for the M1 and M2 sites from the spectra in figure 10.9 are summarized in table 10.4. By comparison with Mössbauer spectra of other silicates, it is assumed that the spectral component with the smaller quadrupole splitting corresponds to Fe^{2+} in the M2 octahedral sites. The distribution of Fe^{2+} in the M1 and M2 sites in the 12018 olivine can be determined on the basis of the relative areas of the M1 and M2 components of the spectrum. Since the lower velocity peaks of the M1 and M2 components of the 12018 olivine spectrum are unresolved, the relative areas of the higher velocity M1 and M2 peaks, labelled B_1 and B_2 in figure 10.9, are taken as a measure of distribution of the Fe^{2+} ions in these two sites. The area of the B_1 peak as a fraction of the total area of B_1 and B_2 is given as $B_1/(B_1 + B_2)$ in table 10.4. A parameter, k, describing the occupancy of the Fe^{2+} ions on the M1 and M2 sites can be defined as

$$k = \frac{(\mathrm{Mg}^{2+}/\mathrm{Fe}^{2+})_{\mathrm{M1}}}{(\mathrm{Mg}^{2+}/\mathrm{Fe}^{2+})_{\mathrm{M2}}}. \qquad (10.1)$$

Here the Fe^{2+} concentrations on the two sites are taken to be

$$(\mathrm{Fe}^{2+})_{\mathrm{M1}} = 2x\frac{B_1}{B_1 + B_2}$$

and

$$(\mathrm{Fe}^{2+})_{\mathrm{M2}} = 2x\frac{B_2}{B_1 + B_2}.$$

If the Fe^{2+} and Mn^{2+} ions are randomly distributed on the M1 and M2 sites, then the parameter k in equation (10.1) would have a value of 1.0. The value of $k = 0.58$ for the 12018 olivine as given in table 10.4, shows a partial ordering of the ions on the sites where Fe^{2+} shows a preference for the M1 sites.

10.4.3 *In situ* studies of Martian samples

The measurement of Mössbauer spectra on Mars presents many challenges. In order to be transported to Mars by spacecraft, a Mössbauer spectrometer must be compact and light. The spectrometer designed for this purpose, MIMOS II, is shown in figure 10.11. Typical laboratory Mössbauer effect experiments utilize a transmission geometry where the absorber is placed between the source and the detector, as discussed in section 8.8. In this case the sample (absorber) is a powder or a thin foil or slice of the material to be studied, typically weighing in the order of 100 mg. This geometry is not suitable for the Mars Mössbauer spectrometer as it would require additional instrumentation to collect and process soil or rock samples to form the absorber. Mössbauer effect experiments can also utilize a backscatter geometry where gamma rays from the source are resonantly absorbed by the sample and the gamma rays that are re-emitted back towards the source are detected by the detector, as shown in figure 8.12. In this way the spectrometer can sample the surface of a bulk material and the spectrum of the resonantly re-emitted gamma rays will show emission peaks rather than absorption peaks.

The Mars Mössbauer spectrometer uses four 1 cm × 1 cm Si-PIN photodiodes for detecting the gamma rays, rather than the more typical proportional counter or scintillation detector, in order to minimize size and weight. As discussed in section 8.8, the use of multiple radiation detectors is effective in optimizing count rate. The overall size of the spectrometer sensor head is about 5 cm × 5 cm × 10 cm and has a circular field of view with a diameter of 3.2 cm. Calibration spectra were recorded

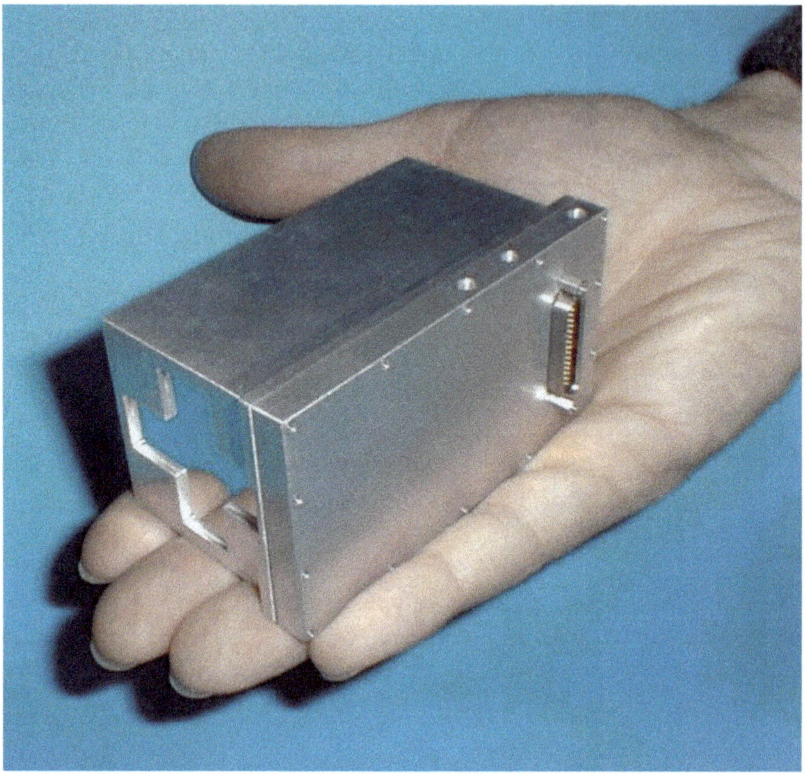

Figure 10.11. ^{57}Fe Mössbauer effect spectrometer MIMOS II used on the NASA Mars Exploration Rovers Spirit and Opportunity. The four 1 cm × 1 cm Si-PIN photodiodes surround the Mössbauer source and are located in the corners of the cross on the left end of the spectrometer, as seen in the image. Courtesy of NASA (2004a). Image credit: NASA/JPL/University of Mainz.

simultaneously in transmission mode using a second (weaker) ^{57}Co source mounted on the back end of the velocity transducer along with an iron calibration foil and detector.

The two NASA spacecraft carrying the rovers Opportunity and Spirit, with Mössbauer spectrometers on board, landed on Mars in January 2004. While they were initially designed to operate for three months, Spirit continued providing data until 2010 and Opportunity was operational until June 2018.

Figure 10.12 shows a Mössbauer spectrum of a sample designated 'Clovis' from the Columbia Hills obtained by the NASA Mars Exploration Rover Spirit. Spectral components can be compared with results from the spectra of terrestrial minerals, and this provides a means of identifying mineral species present on Mars. The spectrum shows the presence of a substantial quantity of goethite, which is an iron-bearing mineral which contains water in the form of hydroxyl. This observation provides convincing evidence of past water activity in the Columbia Hills region of Mars. Spectra of other Martian samples from both Spirit and Opportunity have

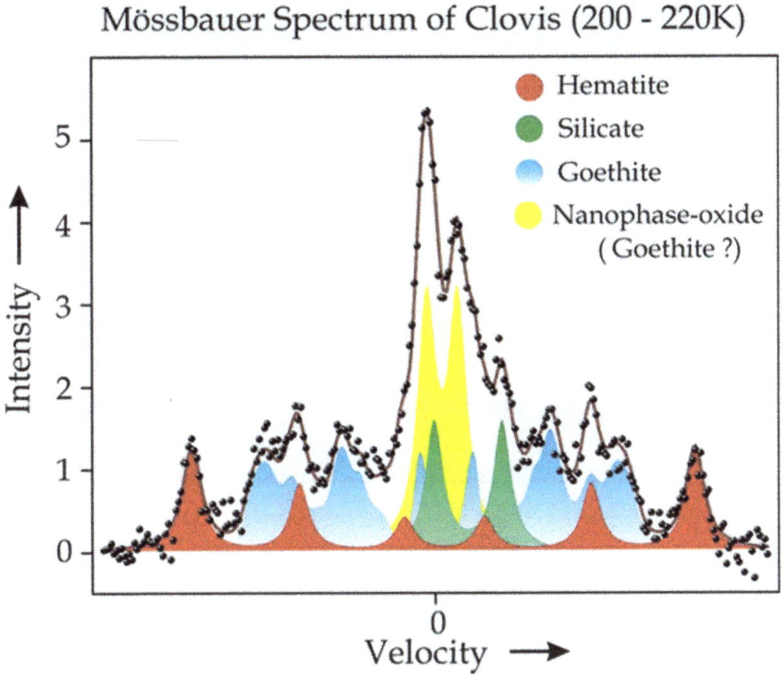

Figure 10.12. ^{57}Fe Mössbauer effect spectrum obtained by MIMOS II on board the NASA Mars Exploration Rover Spirit of 'Clovis', a Martian soil sample from the Columbia Hills. Courtesy of NASA (2004b). Image credit: NASA/JPL/University of Mainz.

shown the presence of other Fe-containing minerals that indicate the presence of an aqueous environment at the time of their formation.

Figure 10.13 shows some Mössbauer results from the NASA Mars Exploration Rover Opportunity. The figure shows spectra of an anomalous rock (designated 'Bounce'), which was located near Opportunity's landing region at Meridiani Planum, along with three spectra of soil samples taken nearby. The three soil samples all show the presence of a Fe^{3+} component in their spectra. The 'Bounce' rock, on the other hand, is lacking any evidence of Fe^{3+} in its Mössbauer spectrum. In fact, 'Bounce's' spectrum is unique to the samples studied by Opportunity and Spirit. This feature is taken as evidence that the 'Bounce' rock did not originate in the plains of Meridiani Planum. For comparison, figure 10.14 shows the Mössbauer spectrum of 'Bounce' along with the spectrum of the Shergotty meteorite. The Shergotty meteorite is a 5 kg meteorite which fell to Earth at Sherghati, in the Gaya district, Bihar, India on 25 August 1865. Based on its chemical composition, it has traditionally been assumed that Shergotty originated on Mars. The spectra shown in figure 10.14, show the similarities between the composition of 'Bounce' and Shergotty. This suggests that Shergotty is, in fact, from Mars and that the composition of 'Bounce' is not unique elsewhere on the planet.

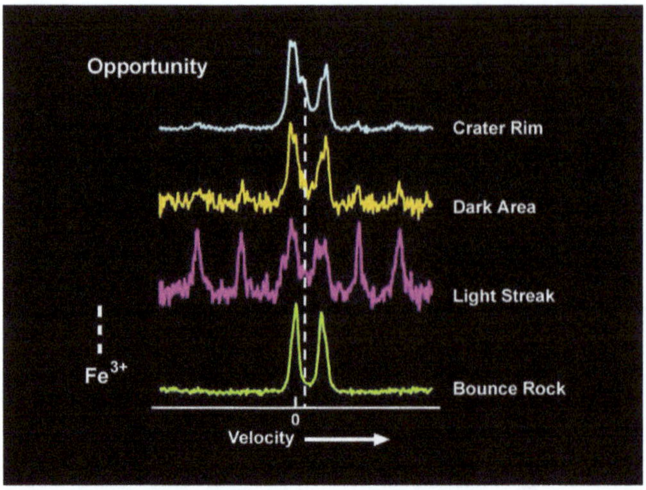

Figure 10.13. ^{57}Fe Mössbauer spectra of 'Bounce' and three soil samples outside of 'Eagle Crater' as obtained by the NASA Mars Exploration Rover Opportunity. https://photojournal.jpl.nasa.gov/catalog/PIA05750. Courtesy of NASA (2004c). Image credit: NASA/JPL/University of Mainz.

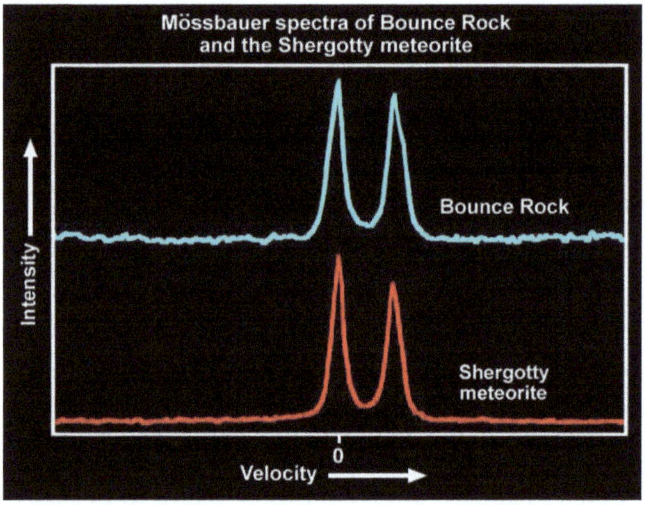

Figure 10.14. Comparison of ^{57}Fe Mössbauer spectra for 'Bounce' obtained by the NASA Mars Exploration Rover Opportunity and the Shergotty meteorite, as described in the text. Courtesy of NASA (2004d). Image credit: NASA/JPL/University of Mainz.

10.5 Archaeological studies

Mössbauer effect spectroscopy has been a useful tool for the study of archaeological materials. Some aspects of the Mössbauer effect that have been utilized are its ability to determine the crystallographic phases present, the distribution of oxidation states and the depth dependence of material properties below the surface of a sample. An overview of the application of Mössbauer spectroscopy to archaeology has been provided by Wagner and Kyek (2004). In the present section we describe three examples of such investigations. These are the study of obsidians, the study of ancient pottery and the study of ancient gold coins.

10.5.1 Obsidians

Obsidian was an important material for tool production by ancient humans. The ability to determine the provenance of obsidian tools in relationship to the location of their recovery is an important indicator of prehistoric trade routes. A number of experimental techniques are available for the characterization of obsidian including x-ray diffraction, x-ray fluorescence, electron spin resonance, neutron activation analysis, thermoluminescence and atomic absorption spectroscopy. As Mössbauer spectroscopy is sensitive to the properties of obsidian, such as composition and oxidation state, it can, in principle, contribute to the classification of these materials according to their source, and a number of investigations of the Mössbauer effect for this purpose have appeared in the literature. A typical example of the results of such an investigation is described in this section.

Longworth and Warren (1979) have undertaken a Mössbauer effect investigation of obsidians from the western Mediterranean region. It is commonly believed that during the period from the Neolithic to the Copper Age, only four sources of obsidian were available in this region: Lipari Island, Pantelleria, the Pontine Islands and Sardinia. Figure 10.15 shows room temperature ^{57}Fe Mössbauer spectra of samples of obsidian from these four locations. Spectra are illustrated for archaeological samples obtained from ancient artifacts and geological samples collected from the location. Spectra were fit to two quadrupole split doublets corresponding to Fe^{2+} and Fe^{3+}. The Fe^{2+} doublets had isomer shifts and quadrupole splittings of ~1.0 and ~1.9 mm s^{-1}, respectively. The Fe^{3+} doublets had isomer shifts and quadrupole splittings of ~0 and 0.3 mm s^{-1}, respectively. As well, some spectra included a sextet associated with Fe_3O_4.

In order to establish a means of identifying the source of an obsidian sample on the basis of its Mössbauer effect spectral parameters, the Fe^{2+} quadrupole splitting has been plotted as a function of the Fe^{3+} quadrupole splitting in figure 10.16. It is clear from the diagram that the parameters for the samples from Pantellaria and from Sardinia each form a reasonably well isolated cluster of points on the graph. The parameters from the Lipari and Pontine samples do not form well defined clusters. Based on this assessment, obsidian samples from Pantellaria or Sardinia can be identified with some degree of certainty on the basis of their Fe^{2+} and Fe^{3+} quadrupole splittings. Samples from Lipari and Pontine, however, cannot be distinguished. Similar plots of other Mössbauer hyperfine parameters, such as isomer shifts, Fe^{3+}/Fe^{2+} area ratios, concentration of Fe_3O_4, show similar ability to identify obsidian source locations. Other studies of the Mössbauer parameters of

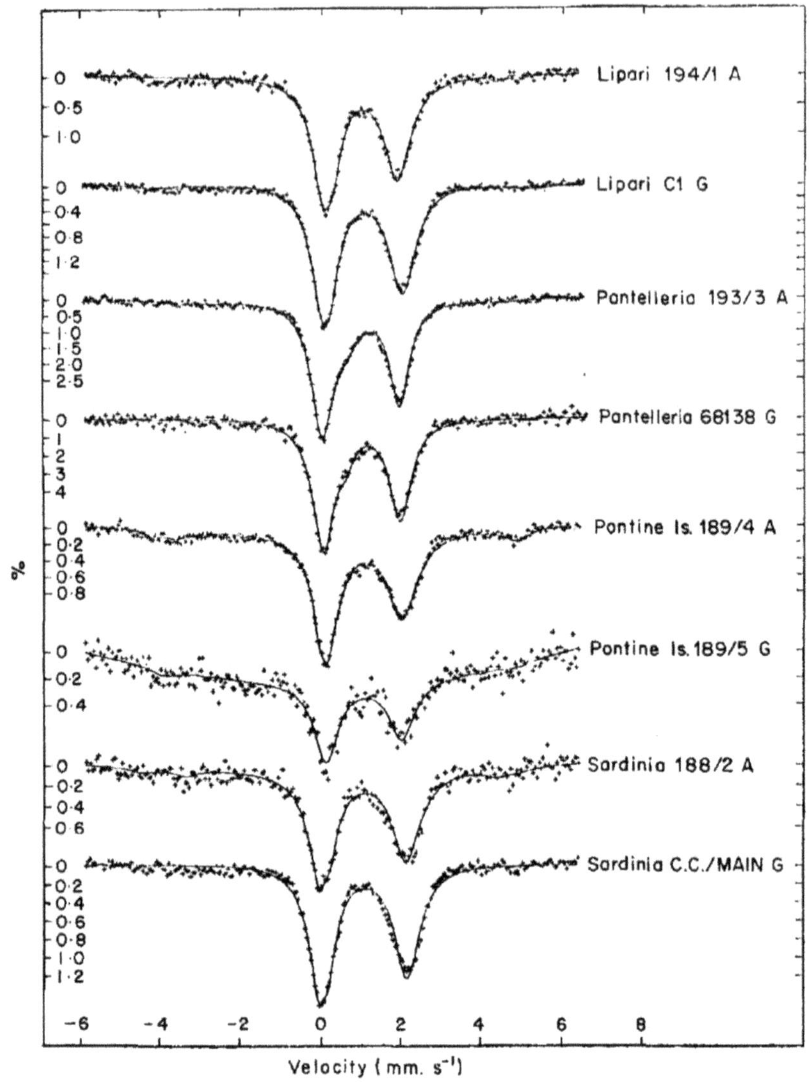

Figure 10.15. Room temperature ^{57}Fe Mössbauer effect spectra of obsidian samples from the four western Mediterranean locations as described in the text. Samples labelled A and G are archaeological and geological samples, respectively. Solid lines are computer fits to the data as described in the text. The velocity scale is referenced to room temperature α-Fe. Reprinted from Longworth and Warren (1979). Copyright (1979). With permission from Elsevier.

obsidians for the purpose of identifying provenance have shown similar results (e.g., Stewart *et al* 2003, Uhm *et al* 2020).

Further Mössbauer studies of obsidians may help to identify an approach for reliably distinguishing between different material sources. However, investigations as discussed here are helpful in conjunction with other experimental techniques, as noted above, in confirming provenance.

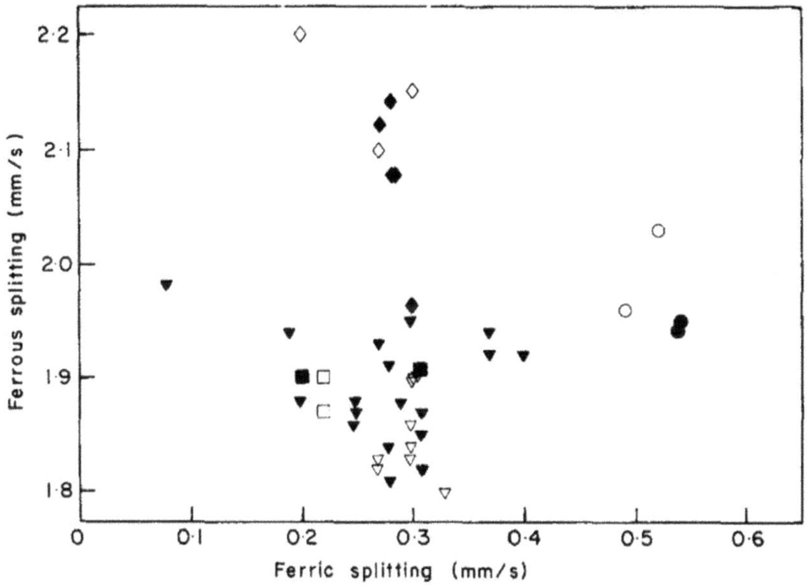

Figure 10.16. Fe^{2+} (ferrous Fe) quadrupole splitting plotted as a function of the Fe^{3+} (ferric Fe) quadrupole splitting. The sample locations are ∇Lipari, ○Pantellaria, □ Pontine and ◇Sardinia. Geological samples are indicated by the solid symbols and the archaeological samples are indicated by the open symbols. Reprinted from Longworth and Warren (1979). Copyright (1979). With permission from Elsevier.

10.5.2 Pottery

The investigation of pottery by Mössbauer spectroscopy typically depends on the presence of iron in clays (e.g., Wagner and Wagner 2004) and investigations have focused on two possible topics, (1) an understanding of the firing conditions and (2) the provenance of the pottery. An example of the application of ^{57}Fe Mössbauer spectroscopy to the study of ancient pottery is given below.

^{57}Fe Mössbauer effect spectra of Celtic pottery samples from the 2nd century BCE in the town of Bopfingen in Baden-Württemberg, Germany have been reported by Friederich *et al* (2010). Figure 10.17 shows typical ^{57}Fe Mössbauer effect spectra obtained from ground sherds of these samples at room temperature and at 4.2 K.

The room temperature Mössbauer spectra in figure 10.17 were fit to three components:

1. A Fe^{3+} quadrupole split doublet, which is the dominant component, with an isomer shift of around +0.44 mm s^{-1} (relative to room temperature α-Fe) and a quadrupole splitting of 0.7–0.8 mm s^{-1}.
2. A weak Fe^{2+} quadrupole split doublet (or in some cases, two Fe^{2+} doublets) with isomer shifts around +1.13 mm s^{-1} and quadrupole splittings of around 2.6 mm s^{-1}.
3. A weak Zeeman split sextet with a hyperfine field between 48.0 and 49.5 T, corresponding to a mixture of hematite and maghemite.

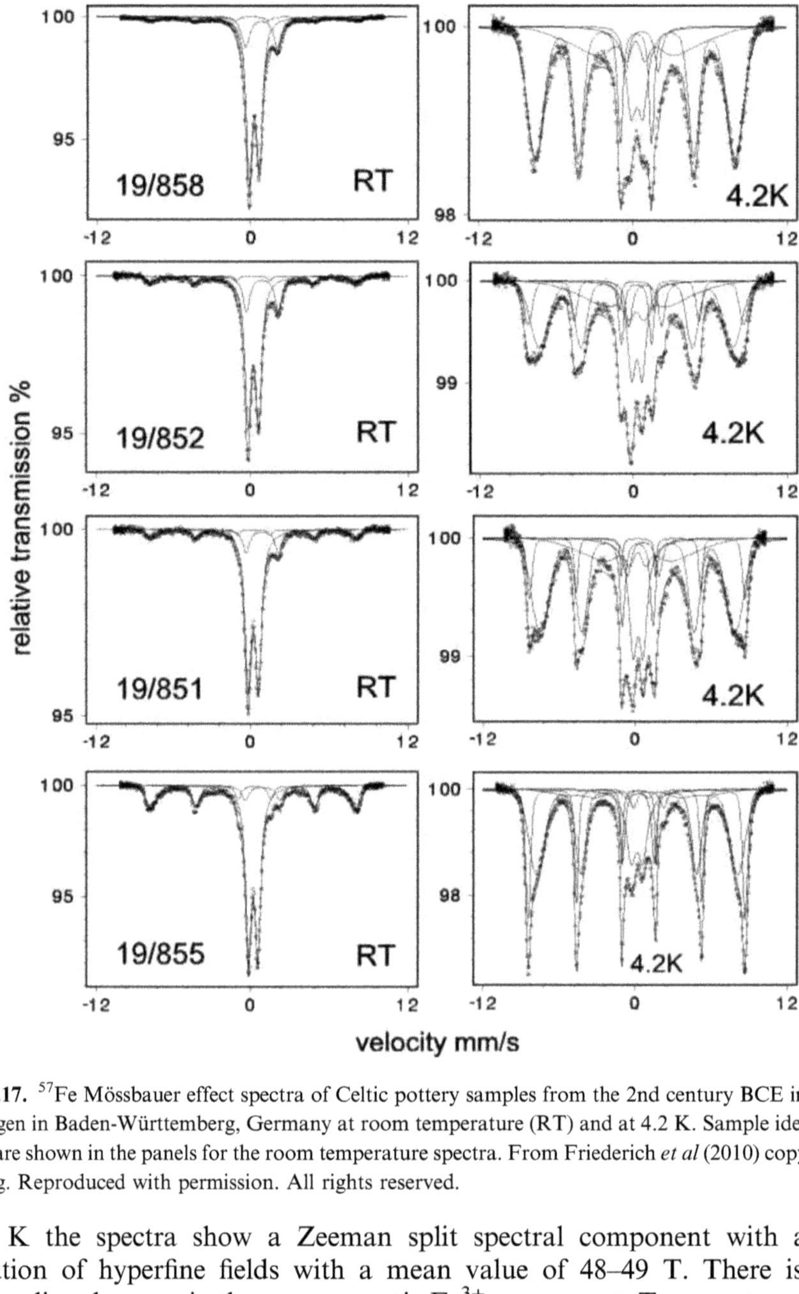

Figure 10.17. ^{57}Fe Mössbauer effect spectra of Celtic pottery samples from the 2nd century BCE in the town of Bopfingen in Baden-Württemberg, Germany at room temperature (RT) and at 4.2 K. Sample identification numbers are shown in the panels for the room temperature spectra. From Friederich *et al* (2010) copyright IOP Publishing. Reproduced with permission. All rights reserved.

At 4.2 K the spectra show a Zeeman split spectral component with a broad distribution of hyperfine fields with a mean value of 48–49 T. There is also a corresponding decrease in the paramagnetic Fe^{3+} component. Temperature dependent studies show that as the sample is cooled the magnetically split component increases in intensity below about 70 K.

Studies of pottery samples that were subsequently heated in the laboratory to various temperatures provided insight into the distribution of magnetic phases as a function of firing temperature and atmosphere and the subsequent cooling process.

Results show that the firing conditions used for early Celtic pottery were fairly consistent over a wide region of Europe. The details of the magnetically split components of the spectra suggest that the pottery was fired at between 800 °C and 850 °C in a reducing environment followed by re-oxidation during cooling.

10.5.3 Metals

^{57}Fe Mössbauer spectroscopy has been utilized to study rust on ancient iron artifacts to understand the effects of environmental conditions (Cindra Fonseca *et al* 2007) and chemical phases and microstructure of ancient weapons (Wagner and Kyek 2004). As well, ^{57}Fe Mössbauer techniques have been used to investigate prehistoric slags from copper, tin and iron production to understand the details of the smelting process (Rüffler 1991, 1996, Rüffler and Gonser 1992). ^{197}Au Mössbauer spectroscopy has also found applications to the study of archaeological samples and here its use for the investigation of ancient gold coins is described.

Celtic gold coins from the La Tène age (480 to 15 BCE) have been found in southern Germany (Morteani and Martinek 1997) and represent an interesting example of ancient metallurgical processes (Gebhard *et al* 1995, Ziegaus 1995). While x-ray fluorescence studies of these coins show a fairly high gold content, electron microprobe studies of cross-sections of cut coins show that the interior of the coin consists of a Au–Ag–Cu alloy with up to 30% Ag–Cu and a surface layer of higher gold content (Lehrberger and Raub 1995). ^{197}Au Mössbauer effect spectroscopy offers a useful non-destructive approach to the investigation of the composition and structure of these coins.

As described in chapter 2, ^{197}Au is a useful Mössbauer nuclide and some of its relevant properties are given in table 2.3. The 18 h half-life parent nuclide ^{197}Pt is prepared via the (n,γ) reaction by thermal neutron irradiation of platinum enriched in ^{196}Pt. The decay of ^{197}Pt by β^- decay is shown in figure 10.18. The $3/2^+ \rightarrow 1/2^+$ 77.3 keV

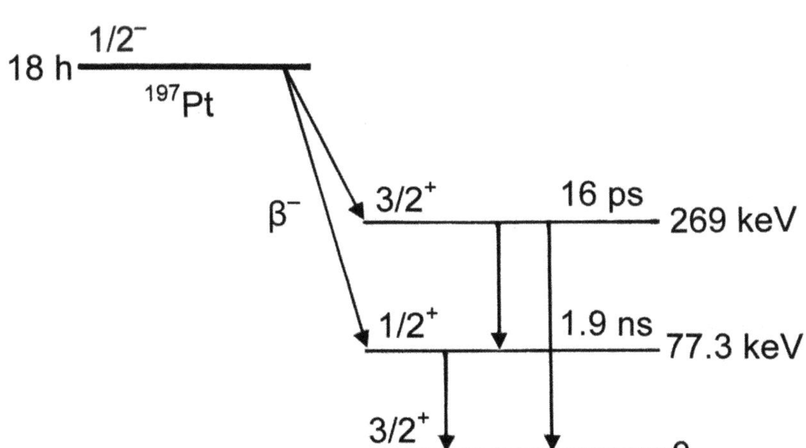

Figure 10.18. Energy level diagram of the 77.3 keV Mössbauer transition in ^{197}Au showing the β^- decay of the precursor ^{197}Pt.

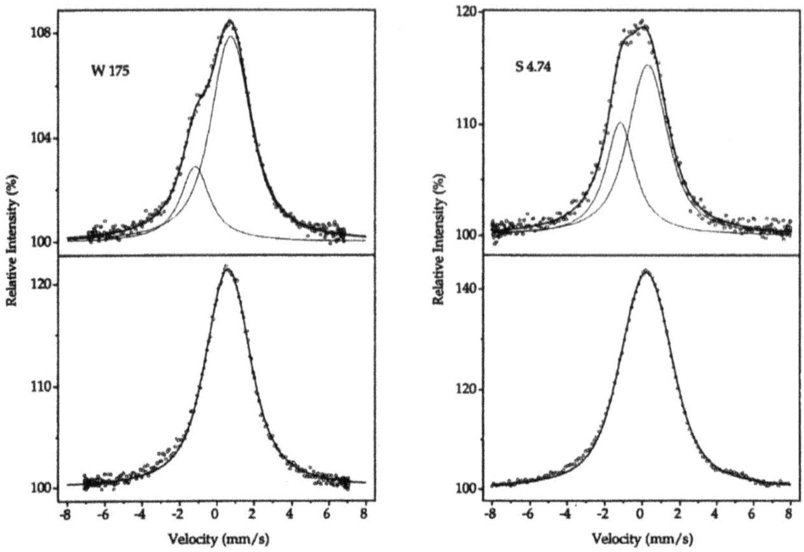

Figure 10.19. ^{197}Au Mössbauer spectra of two ancient Celtic coins, W175 (left) and S4.74 (right) obtained at 4.2 K. Conversion electron spectra are shown on the top and γ-ray backscatter spectra are shown in the bottom. The velocity scale is calibrated relative to the source. From Wagner and Kyek (2004). With permission of Springer.

Mössbauer transition is shown in the figure. The use of backscatter Mössbauer spectroscopy allows for non-destructive analysis and the ability to observe both Mössbauer γ-rays, as well as internal conversion electrons (e.g., Sawicki *et al* 1983) can provide information about the depth dependence of material properties (see sections 8.9 and 9.7). In the case of ^{197}Au, the conversion electrons come from the L_I shell and have an energy of 63 keV (77.3 keV γ-energy minus 14.3 keV binding energy). The penetration depth of these electrons in gold is about 2 μm, compared to about 200 μm for the Mössbauer γ-rays.

Figure 10.19 shows backscatter Mössbauer spectra of two coins (designated W175 and S4.74) obtained at 4.2 K with conversion electrons (top panels) and γ-rays (bottom panels). γ-ray spectra were obtained using an intrinsic germanium detector, while the conversion electron spectra were obtained with a gas-filled detector. Spectra of both coins show similar features. The γ-ray backscatter spectra show a single broad peak near zero velocity, relative to the ^{197}Pt source. Huray *et al* (1971) have shown that the ^{197}Au Mössbauer isomer shift of Au–Cu alloys and Au–Ag alloys exhibit a well defined dependence on composition. For Au–Ag–Cu alloys, isomer shift data may be combined with density measurements to establish both the Ag and Cu content of a Au–Ag–Cu alloy sample.

The conversion electron spectra show both the broad peak near zero velocity, as well as an additional peak with an isomer shift of around -1.23 mm s^{-1}. This velocity is characteristic of the isomer shift of pure Au relative to the ^{197}Pt source and is consistent with the presence of elemental gold on the surface of the coins. The presence of both alloy and pure gold peaks in the conversion electron spectra indicates that this technique is observing both phases. Either the layer of pure gold on the surface is less than the 2 μm penetration depth of the conversion electrons or there are also regions of alloyed gold on the surface.

These results show that ^{197}Au Mössbauer spectroscopy is an effective non-destructive approach to observing the composition and structure of ancient gold coins. This method can effectively complement x-ray fluorescence and electron microprobe studies.

10.6 Biological studies

Iron occurs as a component of many naturally occurring proteins. The existence of iron in these proteins opens up the possibility of using Mössbauer spectroscopy as a means of investigating their properties (see e.g., Kappler *et al* 1967, Moshkovskii 1968, Münck and Bominaar 2012, Schünemann 2021). Unfortunately, the low concentration of iron in most of these materials, combined with their low Debye temperature, make such measurements challenging. Although reducing the sample temperature provides some benefits, obtaining meaningful results in virtually all cases requires the study of samples enriched in ^{57}Fe.

One of the most important and interesting iron-containing proteins is hemoglobin. Hemoglobin is a constituent of red blood cells (i.e., erythrocytes) and serves as an oxygen carrier in many animal species. The utility of hemoglobin as a means of distributing oxygen to the body comes from its ability to combine reversibly with oxygen. Oxyhemoglobin is carried away from the heart by arteries, while reduced hemoglobin is returned to the heart by veins. The Mössbauer effect study of arterial and venous blood, therefore, provides a means of investigating the bonding of oxygen to the iron in the hemoglobin. Hemoglobin also binds to many other molecules and provides a means of transporting these molecules throughout the body. It also binds to carbon dioxide, which inhibits its ability to bind to oxygen and accounts for the resulting toxic effects.

The structure of hemoglobin is illustrated in figure 10.20. The structure shown in the figure is typical of the hemoglobin found in most vertebrates and consists of four globular protein subunits. Each globular subunit has a protein chain bonded to an iron-containing heme group. The heme group, as shown in figure 10.21, consists of a central Fe ion bonded in a planar ring to four neighboring nitrogen atoms. An O_2 molecule can bind covalently to the iron ion to complete an octahedral group as discussed below.

An analysis of the hemoglobin structure shows that it contains about 0.3% Fe, which is insufficient for traditional ^{57}Fe Mössbauer measurements. Most Mössbauer studies of hemoglobin or other proteins have utilized samples enriched in ^{57}Fe, in order to obtain sufficient resonant absorption. One of the earliest studies along these lines by Lang and Marshall (1966a, 1966b), serves as an example of an interesting approach to this problem.

Albino rats were made anemic by the elimination of iron from their diets (Lang and Marshall 1966a, 1966b). They were periodically given injections of iron citrate

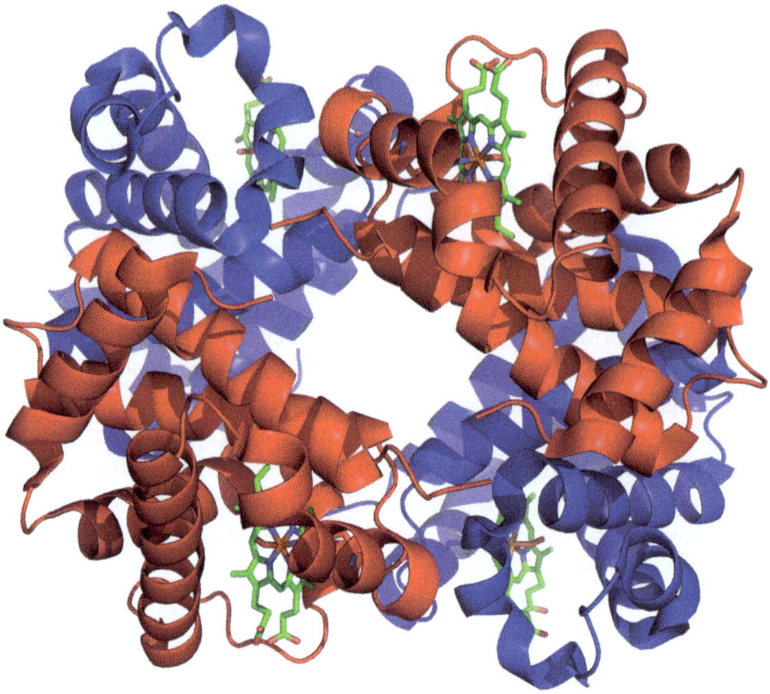

Figure 10.20. The structure of human hemoglobin. The globin subunits are shown in blue and red and the iron-containing heme groups are shown in green. This [Structure of haemoglobin] image has been obtained by the author from the Wikimedia website where it was made available by Zephyris (2007) under a CC BY-SA 3.0 licence. It is included within this chapter on that basis. It is attributed to Zephyris.

enriched to about 80% in ^{57}Fe. These rats provided samples of ^{57}Fe enriched reduced hemoglobin and oxyhemoglobin. ^{57}Fe Mössbauer effect spectra were obtained at various temperatures and typical results are illustrated in figures 10.22 and 10.23. Mössbauer parameters from fits to reduced hemoglobin and oxyhemoglobin as obtained at different temperatures are summarized in table 10.5.

Magnetic susceptibility measurements of reduced hemoglobin (see Lang and Marshall (1966b) and references therein) have shown that the iron is in a high-spin $S = 2$ state. The Mössbauer parameters, as indicated in table 10.5, are consistent with expectations for a high-spin Fe^{2+} state. These values can be compared with those of high-spin silicate minerals, as shown in figure 10.1. The large quadrupole splitting is consistent with an anisotropic Fe environment with planar nitrogen coordination in the heme structure.

The results for oxyhemoglobin are somewhat less definitive than those of reduced hemoglobin. The large quadrupole splitting is indicative of an asymmetric local environment. This indicates that the octahedral nearest neighbor configuration is highly distorted. Further studies have shown that one oxygen atom is covalently bonded to the iron atom and the second oxygen is bonded to the first oxygen at an

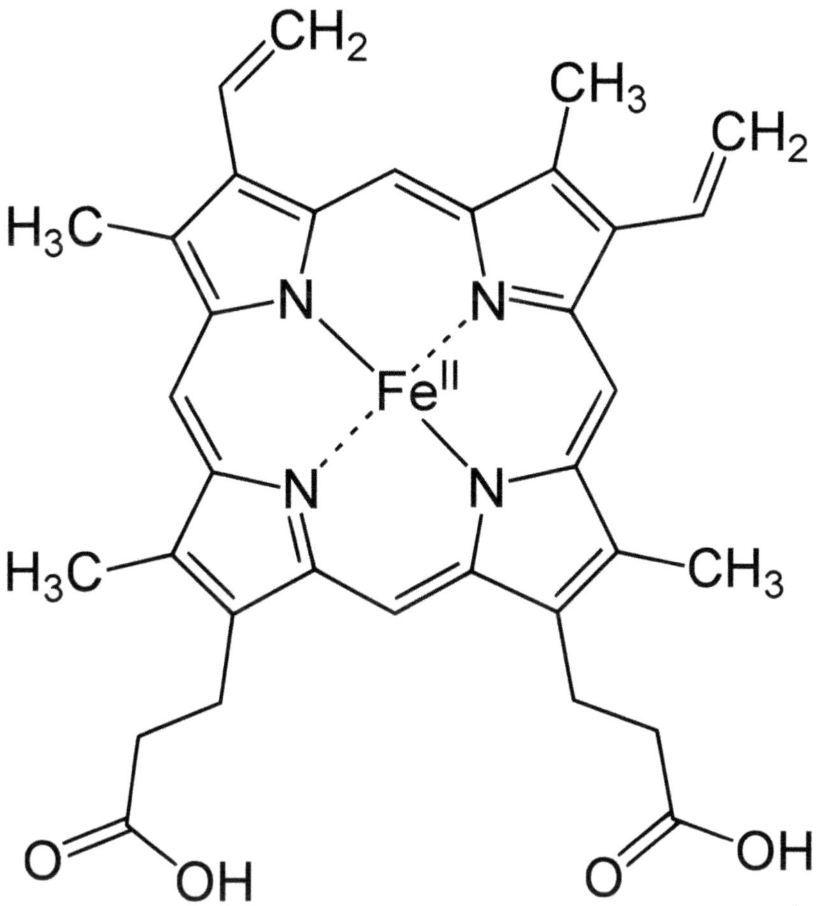

Figure 10.21. The heme B structure. This [Heme B Structure] image has been obtained by the author from the Wikimedia website https://commons.wikimedia.org/wiki/File:Heme_b.svg (Yikrazuul 2010), where it is stated to have been released into the public domain. It is included within this chapter on that basis.

angle. This arrangement is illustrated in figure 10.24. Magnetic susceptibility measurements show that the iron in oxyhemoglobin is in the low-spin $S = 0$ state (Lang and Marshall 1966a). In low-spin Fe^{2+} and Fe^{3+} compounds, ^{57}Fe isomer shifts all typically fall within the range of −0.5 and −0.2 mm s^{-1}, making the distinction between ferrous (Fe^{2+}) and ferric (Fe^{3+}) compounds difficult (Gonser and Grant 1965). Although in the present case, the large quadrupole splitting is more characteristic of Fe^{2+} than Fe^{3+}, the assignment of a specific Fe valence state in a covalently bonded material may not be particularly relevant because of the difficulty of assigning bonding electrons to specific atoms.

Studies of iron-containing proteins, such as the measurements described above, have been important in understanding the role of iron in biological processes. In

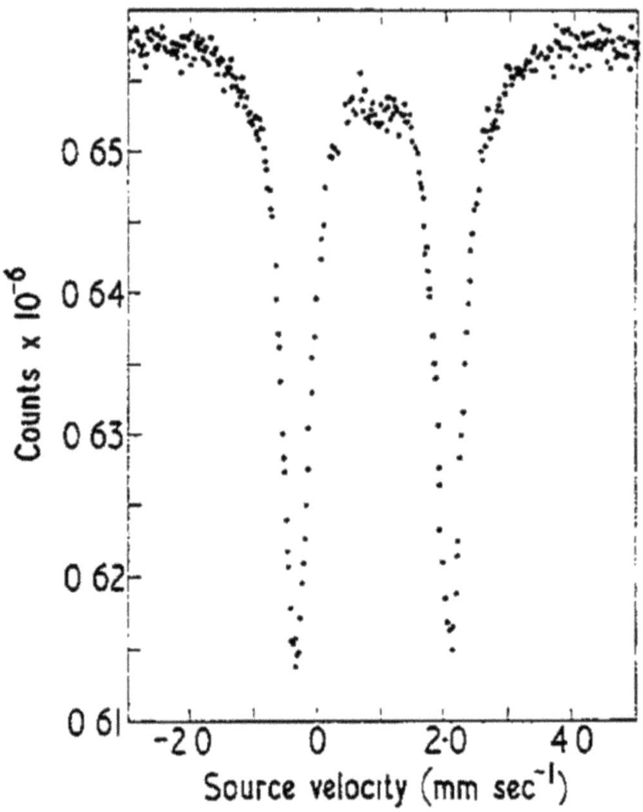

Figure 10.22. ^{57}Fe Mössbauer effect spectra of reduced hemoglobin at 4.2 K. From Lang and Marshall (1966a) copyright IOP Publishing. Reproduced with permission. All rights reserved.

particular, investigations of other molecules bonded to heme proteins have helped clarify the reasons for the physiological effects of these materials. The recent availability of synchrotron-based Mössbauer techniques as described in detail in chapter 11, provides additional opportunities for the investigation of biological materials (see Grünsteudel *et al* 1998).

10.7 Counterfeit currency detection

A particularly novel application of Mössbauer effect spectroscopy is the detection of counterfeit banknotes. The inks used for the printing of genuine banknotes are very specific and are not easily reproduced in counterfeit banknotes. Since genuine printing inks often contain iron, this opens up the possibility of performing Mössbauer effect investigations of suspected currency.

Figure 10.25 shows some typical results of ^{57}Fe Mössbauer effect studies of genuine and counterfeit United States $100 banknotes. The Mössbauer spectrum

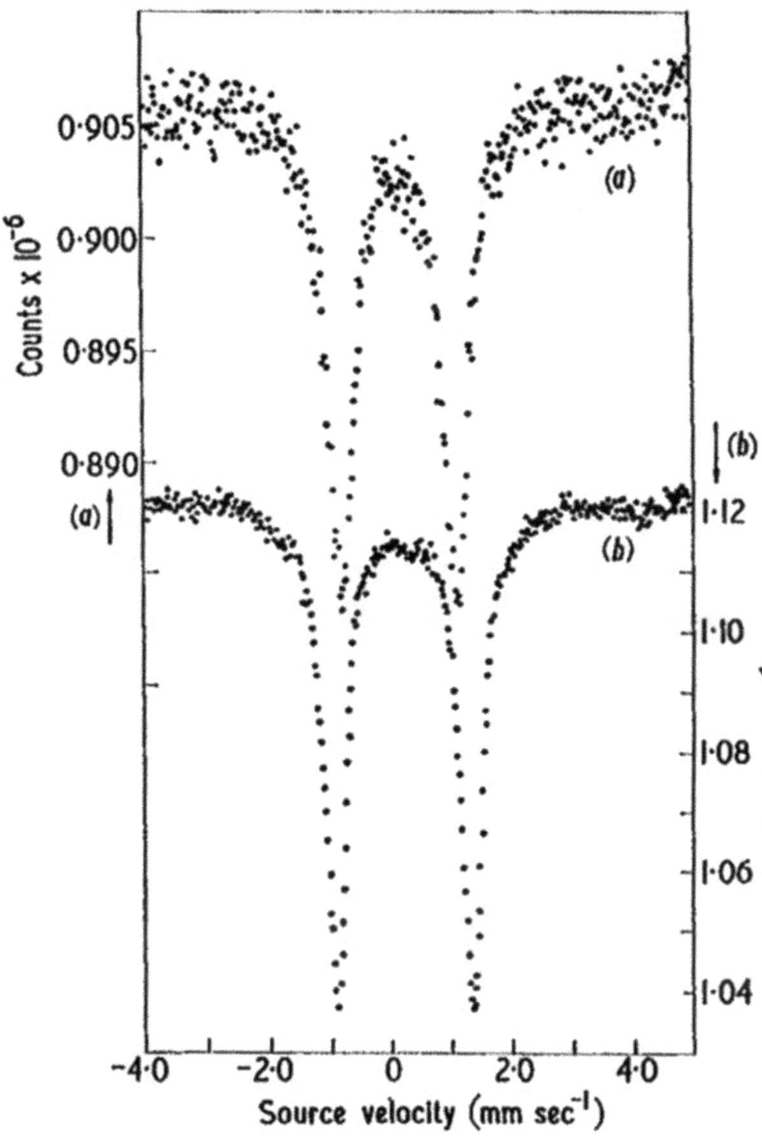

Figure 10.23. ^{57}Fe Mössbauer effect spectra of oxyhemoglobin at (a) 195 K and (b) 1.2 K. From Lang and Marshall (1966a) copyright IOP Publishing. Reproduced with permission.

of the genuine banknote, figure 10.25(a), shows that the black ink used to print the banknote is based on Fe_3O_4 (magnetite) and this spectral component is readily identified on the basis of the results for magnetite (see figure 9.4). The green ink used to print the banknote is based on an Fe-containing chromium oxide and shows a smaller Zeeman splitting than magnetite. The spectrum of the high-quality

Table 10.5. ^{57}Fe Mössbauer effect parameters (isomer shift, δ, and quadrupole splitting, Δ) for reduced hemoglobin (Hb) and oxyhemoglobin (HbO_2) at different temperatures. Values are ± 0.05 mm s^{-1}. Data are adapted from Lang and Marshall (1966a).

Sample	T (K)	δ (mm s^{-1})	Δ (mm s^{-1})
Hb	4	+0.91	2.40
Hb	195	+0.90	2.40
HbO_2	1.2	+0.24	2.24
HbO_2	77	+0.26	2.19
HbO_2	195	+0.20	1.89

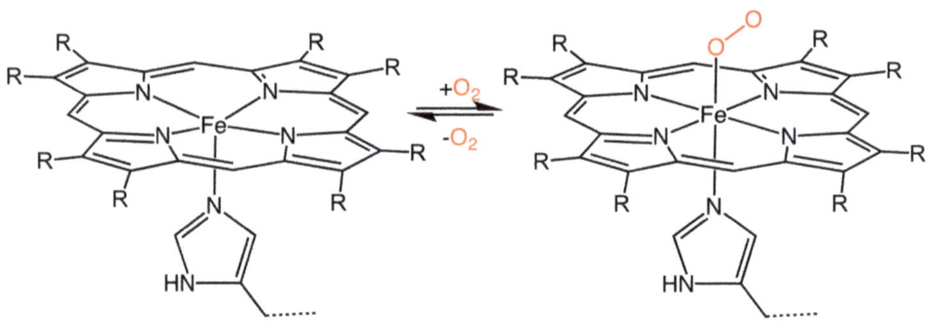

Figure 10.24. Structure of O_2 bonding to heme. This [Oxygenation of heme protein] image has been obtained by the author from the Wikimedia website where it was made available by Smokefoot (2017) under a CC BY-SA 4.0 licence. It is included within this chapter on that basis. It is attributed to Smokefoot.

counterfeit banknote (figure 10.25(b)) is visually very similar to the genuine banknote. However, a close examination will show clear differences and a detailed computer analysis shows that the hyperfine parameters obtained for the counterfeit banknote are not within the range of values that are obtained for genuine banknotes.

Figure 10.25(c) shows the Mössbauer effect spectrum of a low-quality counterfeit banknote. This spectrum is easily distinguished from that of the genuine banknote. The low-quality counterfeit banknote is printed with a carbon-based ink which contains very little iron.

Because of the small mass of ink used on a banknote, a Mössbauer effect spectrum with acceptable statistics may take several days to obtain. Although this may not be an efficient or economical method for the routine investigation of suspect currency, it has the capability of determining if counterfeit banknotes circulated in different locations have a common origin.

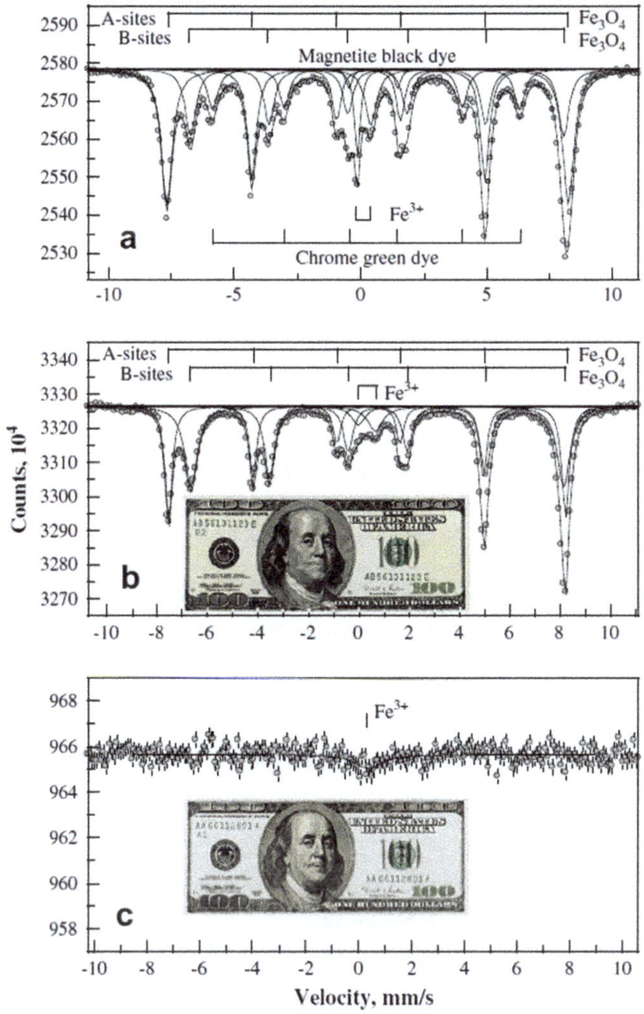

Figure 10.25. ^{57}Fe Mössbauer effect spectra of (a) a genuine United States $100 banknote, (b) high quality counterfeit United States $100 banknote and (c) low quality United States $100 banknote. Note: The 'A' sites are the tetrahedral sites, and the 'B' sites are the octahedral sites, as described above. Reprinted from Rusanov *et al* (2009). Copyright (2009). With permission from Elsevier.

Bibliography and suggestions for further reading

Amthauer G, McIver J R and Viljoen E A 1979 ^{57}Fe and ^{119}Sn Mössbauer studies of natural tin-bearing garnets *Phys. Chem. Minerals* **4** 235–44

Bancroft G M, Maddock A G and Rurns R G 1967 Applications of the Mössbauer effect to silicate mineralogy—I. Iron silicates of known crystal structure *Geochim. Cosmochim. Acta* **31** 2219–46

Barb D, Morariu M, Diamandescu L, Znamirovschi V, Ciurchea D and Motiu A 1982 Mössbauer studies on Moci (Roumania) meteorite *Ann. Geophys.* **38** 875–9

Bayer G, Felsche J, Schulz H and Rüegsegger P 1972 X-ray study and Mössbauer spectroscopy on lunar ilmenites (Apollo 11) *Earth Planet. Sci. Lett.* **16** 273–4

Berry F J, Bland P A, Oates G and Pillinger C T 1994 Fe-57 Mössbauer spectroscopic studies of the weathering of L-chondrite meteorites *Hyperfine Interact.* **91** 577–81

Bonazzi A, Ortalli I, Pedrazzi G, Jiang K and Zhang X 1992 Mössbauer, X-rays and SEM analysis of meteorites from the mineralogy museum of the University of Parma (Italy) *Hyperfine Interact.* **70** 953–6

Cindra Fonseca M P, Bastos I N, Caytuero A and Baggio Saitovitch E M 2007 Rust formed on cannons of XVIII century under two environment conditions *Corros. Sci.* **49** 1949–62

Coey J M D 1980 Clay minerals and their transformations studied with nuclear techniques: the contribution of Mössbauer spectroscopy *At. Energy Rev.* **18** 73–124

Coey J M D 1984 Mössbauer spectroscopy of silicate materials *Mössbauer Spectroscopy Applied to Inorganic Chemistry* ed G I Long (New York: Plenum) pp 443–509

Costa B F O, Alves E I, Silva P A O C and Batista A C 2021 ^{57}Fe Mössbauer analysis of meteorites and tektites *Minerals* **11** 628

Dunlap R A and Sibley A D E 2004 A Mössbauer effect study of Fe-site occupancy in Australasian tektites *J. Non-Cryst. Solids* **337** 36–41

Eibschütz M and Ganiel U 1967 Mössbauer studies of Fe^{2+} in paramagnetic fayalite (Fe_2SiO_4) *Sol. State Commun* **5** 267–70

Friederich J, Gebhard R, Krause R, Riederer J, Wagner F E and Wagner U 2010 Mössbauer study of Celtic pottery from Bopfingen, Baden-Württemberg, Germany *J. Phys.: Conf. Ser.* **217** 012064

Gebhard R, Lehrberger G, Morteani G, Raub C, Wagner F E and Wagner U 1995 Coin moulds and other ceramic material: a key to Celtic precious metal working Prehistoric Gold in Europe: Mines *Metallurgy and Manufacture—Proc. of the NATO Advanced Research Workshop on Prehistoric Gold in Europe (Seeon, Germany, September 27–October 1, 1993) Vol 280 of NATO ISI Series E: Applied Sciences* ed G Morteani and J P Northover (Dordrecht: Kluwer Academic Publishers) pp 273–301

Gonser U and Grant R W 1965 Mössbauer effect in hemoglobin and some iron-containing biological compounds *Biophys. J.* **5** 823–44

Greenwood N N and Gibb T C 1971 *Mössbauer Spectroscopy* (London: Chapman and Hall)

Grünsteudel H *et al* 1998 Mössbauer spectroscopy with synchrotron radiation: a new technique entering biological inorganic chemistry *Inorgan. Chim. Acta* **275–276** 334–9

Huray P G, Roberts L D and Thomson J O 1971 Study of the Cu–Au and Ag–Au alloy systems as a function of composition and order through the use of the Mössbauer effect for ^{197}Au *Phys. Rev.* B **4** 2147–61

Johnson C E, Rideout M S and Cranshaw T E 1963 The Mössbauer effect in iron alloys *Proc. Phys. Soc.* **81** 1079–90

Johnson C E, Rideout M S, Cranshaw T E and Madsen P E 1961 Hyperfine field and atomic moment of iron in ferromagnetic alloys *Phys. Rev. Lett.* **6** 450–1

Kalvius M and Kienle P (ed) 2012 *The Rudolf Mössbauer Story: His Scientific Work and Its Impact on Science and History* (Heidelberg: Springer)

Kappler H M, Trautwein A, Mayer A and Vogel H 1967 Mechanical constant velocity drive for Mössbauer spectroscopy of hemoglobin *Nucl. Instrum. Methods* **53** 157–62

Kyek A, Wagner F E, Lehrberger G, Pankhurst Q A and Ziegaus B 2000 Celtic gold coins in the light of Mössbauer spectroscopy, electron microprobe analysis and X-ray diffraction *Hyperfine Interact.* **126** 235–40

Lang G and Marshall W 1966a Mössbauer effect in some haemoglobin compounds *Proc. Phys. Soc.* **87** 3–34

Lang G and Marshall W 1966b Mössbauer effect in some haemoglobin compounds *J. Mol. Biol.* **18** 385–404

Lehrberger G and Raub C 1995 A look into the interior of Celtic gold coins *Prehistoric Gold in Europe: Mines, Metallurgy and Manufacture Proc. of the NATO Advanced Research Workshop on Prehistoric Gold in Europe (Seeon, Germany, September 27–October 1, 1993) Vol 280 of NATO ISI Series E: Applied Sciences* ed G Morteani and J P Northover (Dordrecht: Kluwer Academic Publishers) pp 341–55

Longworth G and Warren S E 1979 The application of Mössbauer spectroscopy to the characterisation of Western Mediterranean obsidian *J. Archaeol. Sci.* **6** 179–93

Maksimova A A and Oshtrakh M I 2021 Applications of Mössbauer spectroscopy in meteoritical and planetary science, part I: undifferentiated meteorites *Minerals* **11** 612

Molera J, Pradell T, Salvado N and Vendrell-Saz M 1999 Evidence of tin oxide recrystallization in opacified lead glazes *J. Am. Ceram. Soc.* **82** 2871–5

Morteani G and Martinek K P 1997 Die chemische analyse prähistorischer Goldobjekte *Das Prähistorische Gold in Bayern, Böhmen und Mähren* ed G Lehrberger, J Friedrich, R Gebhard and J Hrala (Prague: Institute of Archaeology) pp 127–35

Moshkovskii Y S 1968 Applications of the Mössbauer effect in biology *Chemical Applications of Mössbauer Spectroscopy* ed V I Goldanskii and R H Herber (New York: Academic) ch 10

Münck E and Bominaar E L 2012 Mössbauer spectroscopy of biological systems *The Rudolf Mössbauer Story: His Scientific Work and Its Impact on Science and History* ed M Kalvius and P Kienle (Berlin: Springer) ch 13

NASA 2004a *PIA05111: Powerful Yet Tiny Machine* (https://photojournal.jpl.nasa.gov/catalog/PIA05111)

NASA 2004b *PIA07104: Water-Signature Mineral Found by Spirit* (https://photojournal.jpl.nasa.gov/catalog/PIA07104)

NASA 2004c *PIA05750: 'Bounce': Not Like the Others* (https://photojournal.jpl.nasa.gov/catalog/PIA05750)

NASA 2004d *PIA05751: 'Bounce' and Shergotty Share Common Ground* (https://photojournal.jpl.nasa.gov/catalog/PIA05751)

Perditax 2010 *Crystal Structure of Mg_2SiO_4 (Olivine, PNMA) Projected onto the (a, b) Plane* (https://commons.wikimedia.org/wiki/File:Olivine_Mg2SiO4.png)

Persoons R M, De Grave E and Vandenbergh R E 1990 Mössbauer study of co-substituted magnetite *Hyperfine Interact.* **54** 655–60

Roberts S E, Sheffer A A, McCanta M C, Dyar M D and Sklute E C 2019 Oxidation state of iron in fulgurites and Trinitite: Implications for redox changes during abrupt high-temperature and pressure events *Geochim. Cosmochim. Acta* **266** 332–50

Rossano S, Balan E, Morin G, Bauer J P, Calas G and Brouder C 1999 ^{57}Fe Mössbauer spectroscopy of tektites *Phys. Chem. Min* **26** 530–8

Rüffler R 1991 Three thousand years of copper metallurgy—seen through the eyes of Mössbauer spectroscopy. Part I: possibilities of estimating process parameters (as tested on modern industrial slags) *Hyperfine Interact.* **67** 473–82

Rüffler R 1996 Three thousand years of copper metallurgy—seen through the eyes of Mössbauer spectroscopy. Part III: copper losses in bronze age slags *Hyperfine Interact.* **99** 401–7

Rüffler R and Gonser U 1992 Three thousand years of copper metallurgy—seen through the eyes of Mössbauer spectroscopy. Part II: a bronze age copper process in the Alpine Region *Hyperfine Interact.* **70** 1005–8

Rusanov V, Chakarova K, Winkler H and Trautwein A X 2009 Mössbauer and X-ray fluorescence measurements of authentic and counterfeited banknote pigments *Dyes Pigm.* **81** 254–8

Sawicki J A, Tyliszczak T, Stanek J, Sawicka B D and Kowalski J 1983 Conversion electron Mössbauer spectroscopy of ^{197}Au *Nucl. Instrum. Methods Phys. Res.* **215** 567–8

Schünemann V 2021 From small molecules to complexes systems: a survey of chemical and biological applications of the Mössbauer effect *Modern Mössbauer Spectroscopy: New Challenges Based on Cutting-Edge Techniques* ed Y Yoshida and G Langouche (Singapore: Springer Nature) ch 4

Smokefoot 2017 *Oxygenation of Heme Protein* (https://commons.wikimedia.org/wiki/File:Mboxygenation.png)

Stewart S J, Cernicchiaro G, Scorzelli R B and Poupeau G 2003 Magnetic properties and ^{57}Fe Mössbauer spectroscopy of Mediterranean prehistoric obsidians for provenance studies *J. Non-Cryst. Solids* **323** 188–92

Uhm Y R, Sun G M, Jin M E, Jwa Y J, Seo J Y, Choi H and Kim C S 2020 Provenance studies for prehistoric obsidian by using Mössbauer spectroscopy *J. Korean Phys. Soc.* **77** 253–7

Virgo D and Hafner S S 1972 Temperature-dependent Mg, Fe distribution in a lunar olivine *Earth Planet. Sci. Lett.* **14** 305–12

Wagner F E and Kyek A 2004 Mössbauer spectroscopy in archaeology: introduction and experimental considerations *Hyperfine Interact.* **154** 5–33

Wagner F E and Wagner U 2004 Mössbauer spectra of clays and ceramics *Hyperfine Interact.* **154** 35–82

Yikrazuul 2010 *Heme B Structure* (https://commons.wikimedia.org/wiki/File:Heme_b.svg)

Yoshida Y and Langouche G (ed) 2013 *Mössbauer Spectroscopy: Tutorial Book* (Heidelberg: Springer)

Zephyris 2007 *Structure of Haemoglobin* (https://commons.wikimedia.org/wiki/File:1GZX_Haemoglobin.png)

Zhang Y C, Stevens J G, Li Y S and Li Z L 1994 Mössbauer study of the Jilin and Xinyang meteorites *Hyperfine Interact.* **91** 547–50

Ziegaus B 1995 New aspects on Celtic coin hoards in Southern Germany *Prehistoric Gold in Europe: Mines, Metallurgy and Manufacture—Proc. of the NATO Advanced Research Workshop on Prehistoric Gold in Europe (Seeon, Germany, September 27–October 1, 1993) Vol 280 of NATO ISI Series E: Applied Sciences* ed G Morteani and J P Northover (Dordrecht: Kluwer Academic Publishers) pp 597–608

IOP Publishing

The Mössbauer Effect (Second Edition)

Richard A Dunlap

Chapter 11

Ion implantation and synchrotron radiation-based Mössbauer studies

11.1 Introduction

The Mössbauer effect techniques, as described previously, utilize traditional experimental methods, as described in chapter 8. While there are a substantial number of nuclides that are suitable for traditional Mössbauer studies, there are limitations. In some cases, for example, the concentration of the Mössbauer nuclide is too small to obtain meaningful results with naturally occurring samples or there is no appropriate precursor for the Mössbauer transition. Recent development of techniques that utilize accelerators have helped to expand the scope of Mössbauer spectroscopy. Two important examples, ion implantation and synchrotron radiation-based Mössbauer spectroscopy are discussed in the present chapter.

11.2 Ion implantation studies

Ion implantation is commonly used in industrial processes such as semiconductor device fabrication (see e.g., Armini *et al* 1982). The basic design of an ion implanter is illustrated in figure 11.1. Positive ions are introduced into the ion implanter as shown in the diagram. The basic operation of the ion source is described in Jin *et al* (2021). Ions are accelerated to 50 keV to 100 keV with an appropriate negative voltage. These accelerated ions are passed through a region of magnetic field which acts as a mass separator to choose the desired nuclide for implantation. The final stage can consist of additional acceleration (up to about 500 keV) and a magnetic quadrupole focusing lens.

The implantation depth depends on a number of factors including the ion energy, the ion mass and the composition of the target. Typical implantation profiles are illustrated in figure 11.2. This figure shows implantation depths of up to about 0.2 μm for implantation energies of 100 keV. Implantation depths of the order of a μm or so can be achieved with energies of 500 keV.

doi:10.1088/978-0-7503-6039-5ch11

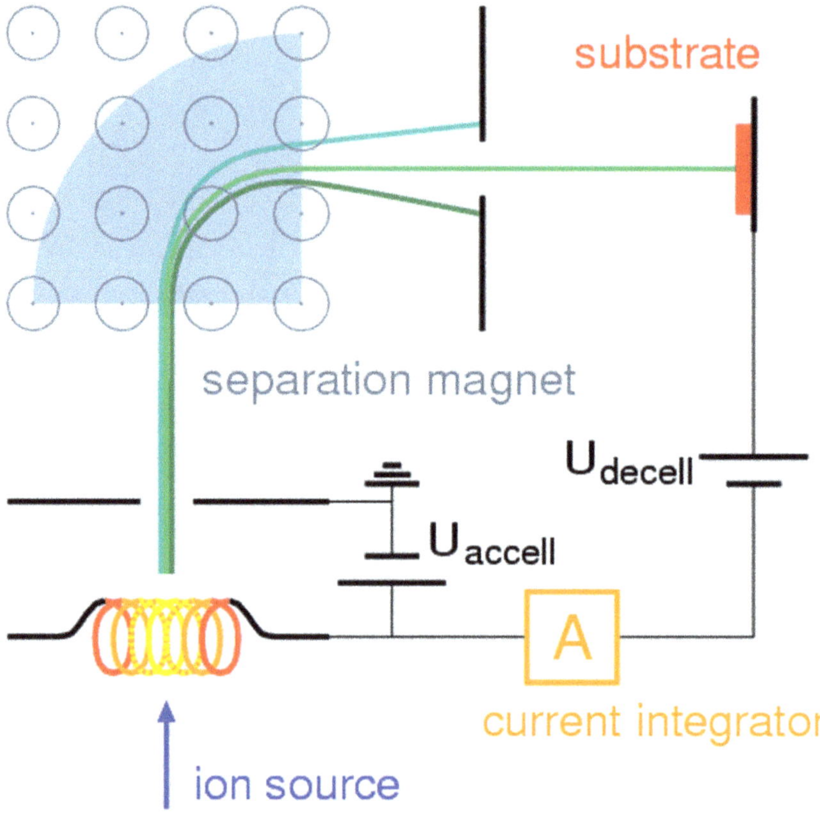

Figure 11.1. Basic design of an ion implanter. This [schematics of an mass separating implantation/deposition setup] image has been obtained by the author from the Wikimedia website where it was made available by Schwen (2005) under a CC BY-SA 3.0 licence. It is included within this chapter on that basis. It is attributed to Daniel Schwen.

The implanted ion loses energy in the target as a result of Coulomb interactions with charges in the material. For energies much larger than a few tens of keV, the interaction of the ion is primarily with electrons in the material and these electrons can be liberated from atoms in the target. For energies in the tens of keV range and less, nuclear interactions between the ion and the target nuclei cause atoms to be displaced from their lattice sites.

The behavior of the damage caused by ion implantation as described above depends on the specifics of the target material. In metals, it is common for damage to completely anneal out at room temperature. In semiconducting materials defect configurations caused by implantation damage are stable at room temperature and annealing at elevated temperatures is necessary for defect recovery. Thus, ion implantation at low temperature and/or subsequent annealing of the implanted sample may form an integral part of the investigation.

A final point concerning the use of ion implantation for Mössbauer studies concerns the implantation of radioactive nuclei that act as precursors for Mössbauer

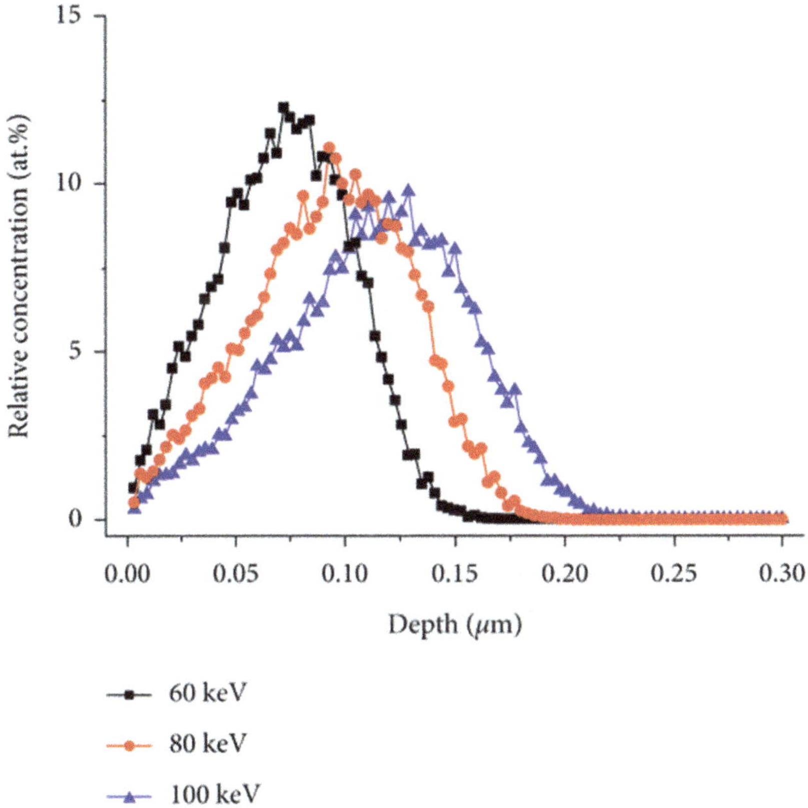

Figure 11.2. Calculated distributions of nitrogen ion distributions in AISI M50 steel as a function of implantation energy. Figure 2 from Xie *et al* (2021) Copyright 2021, CC BY 4.0

effect spectroscopy. Appropriate care must be taken to contain radioactive materials to ensure the safe operation of the implanter. As such, ion implanters that are used for Mössbauer effect studies are typically dedicated facilities.

Ion implantation Mössbauer effect experiments basically fall into three categories, depending on the nature of the implanted ions (Langouche and Yoshida 2013). Specifically, it is the lifetime of the implanted nucleus that categorizes the experiment. Off-line experiments involve Mössbauer parent nuclei with long lifetimes. These, in principle, allow for the creation of the sample in the implanter and subsequent Mössbauer measurements in a traditional experiment. On-line experiments involve Mössbauer parent nuclei with lifetimes that are sufficiently short that the off-line approach is not possible and Mössbauer measurements must be made while the implanted sample remains in the implanter. In-beam experiments are an extreme case of on-line experiments (although the terms are sometimes used interchangeably) where the lifetime of the Mössbauer parent nuclei is comparable to the lifetime of the Mössbauer nuclear state. The next three sections describe these three categories of ion implantation Mössbauer effect experiments with specific reference to how they apply to ^{57}Fe Mössbauer spectroscopy. It should be noted, however, that numerous other Mössbauer transitions can be studied by these methods.

11.3 Off-line Mössbauer spectroscopy

Off-line ^{57}Fe Mössbauer studies utilize samples implanted with ^{57}Fe ions (e.g., Sawicka and Sawicki 1977) or ^{57}Co ions (e.g., Langouche and De Potter 1987). In the case of ^{57}Fe implantation, the technique is useful for the investigation of materials with small iron content or for the study of Fe impurity site hyperfine parameters. This approach is analogous to Mössbauer studies where samples with Fe enriched in ^{57}Fe are prepared by traditional techniques (see section 10.6). In the case of implanted ^{57}Co, this radioactive nuclide is prepared as discussed in section 2.5, typically by deuteron irradiation of ^{56}Fe as given in equation (2.7). The ^{57}Co is then implanted into the sample to be investigated as described above. The ion implanted sample becomes the source of ^{57}Fe Mössbauer γ-rays according to the decay scheme shown in figure 2.8. As the half-life of ^{57}Co is 270 days, the implanted source can be removed from the implanter for use in a traditional Mössbauer spectrometer. However, in many cases, environmental conditions, such as temperature and atmosphere, would be detrimental to the properties of the material under investigation and measurements with the source inside the implanter are preferable. The ^{57}Fe source nuclei, which undergo the 14.4 keV Mössbauer transition are in the material under investigation and the spectrum of the emitted γ-radiation will exhibit the hyperfine features of the Fe sites in the host material. Thus, a single-line absorber containing ^{57}Fe nuclei is used to obtain the Mössbauer spectrum. Typical single-line absorbers are stainless steel or potassium ferrocyanide ($K_4[Fe(CN)_6]\cdot 3H_2O$), often enriched in ^{57}Fe. This approach to the use of ^{57}Co doped material as a Mössbauer source is analogous to traditional experiments referred to as emission Mössbauer or source Mössbauer experiments where radioactive ^{57}Co is diffused into the sample to be investigated by chemical means (see e.g., Jha *et al* 1990, Nagy 1994). As noted above, other long-lived Mössbauer precursor nuclei in addition to ^{57}Co can be implanted into a host material to investigate hyperfine parameters at other elemental sites.

An important application of ion implantation Mössbauer spectroscopy has been the investigation of impurities in semiconductors (e.g., Langouche 1992). One significant question concerning ion implantation in semiconductors is the crystallographic location of the implanted ion. ^{57}Fe off-line Mössbauer spectra of ^{57}Co implanted silicon have been reported by Langouche and De Potter (1987) and typical results are illustrated in figure 11.3. Single crystal Si was implanted at liquid He temperature with ^{57}Co ions at an energy of 80 keV. After implantation the sample served as the source of ^{57}Fe Mössbauer γ-rays and could be studied as a function of temperature. Spectra were recorded using a room temperature moving sodium ferrocyanide absorber.

The differences between the spectra in figure 11.3 as a function of temperature can be explained in terms of the temperature dependence of the mobility of atomic

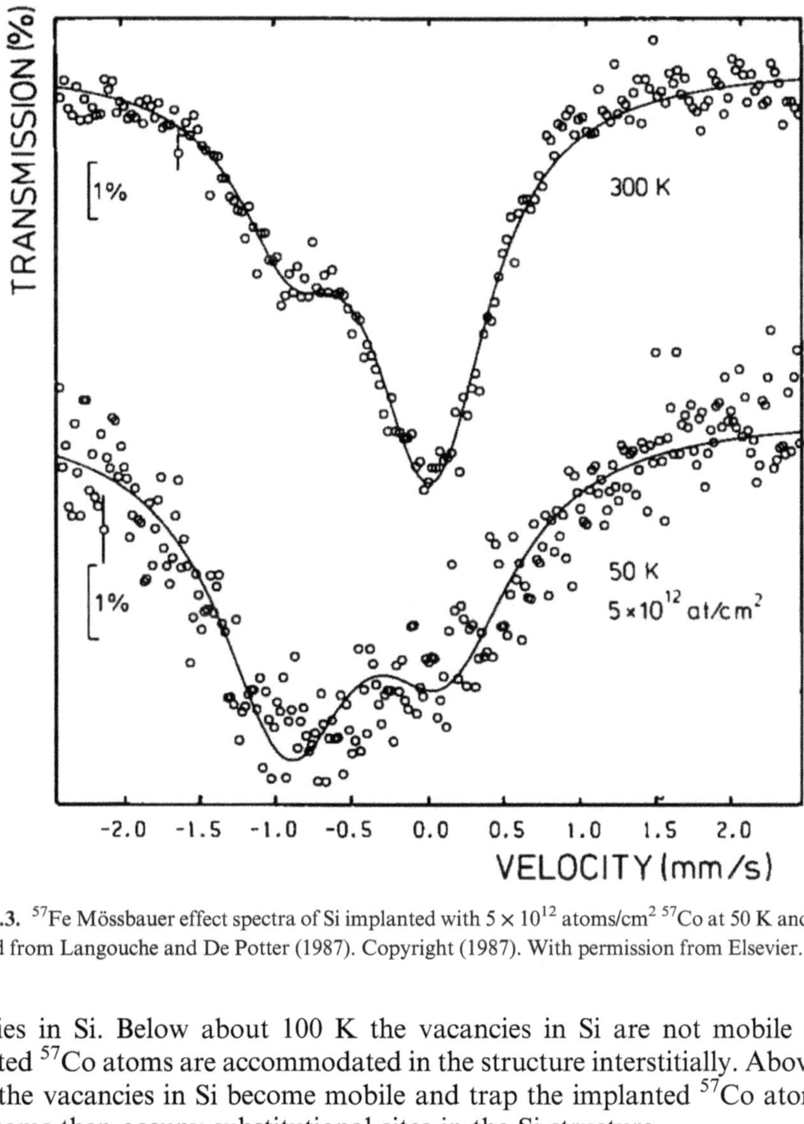

Figure 11.3. ^{57}Fe Mössbauer effect spectra of Si implanted with 5×10^{12} atoms/cm^2 ^{57}Co at 50 K and at 300 K. Reprinted from Langouche and De Potter (1987). Copyright (1987). With permission from Elsevier.

vacancies in Si. Below about 100 K the vacancies in Si are not mobile and the implanted ^{57}Co atoms are accommodated in the structure interstitially. Above about 100 K the vacancies in Si become mobile and trap the implanted ^{57}Co atoms. The ^{57}Co atoms then occupy substitutional sites in the Si structure.

11.4 On-line Mössbauer spectroscopy

Typical on-line ^{57}Fe Mössbauer effect studies use ^{57}Mn ions implanted into the material of interest. The decay mode of ^{57}Mn is illustrated in figure 11.4, showing how the 14.4 keV ^{57}Fe Mössbauer effect level is populated. As the half-life of the ^{57}Mn ground state is about 1.4 min, Mössbauer experiments must be conducted with the ^{57}Mn source material inside the implanter.

Worldwide, two research facilities have undertaken studies involving this technique, ISOLDE (Isotope mass Separator On-Line facility) at CERN (e.g., Weyer 2000) and the RIKEN facility in Tokyo (e.g., Kobayashi *et al* 2000). A diagram of

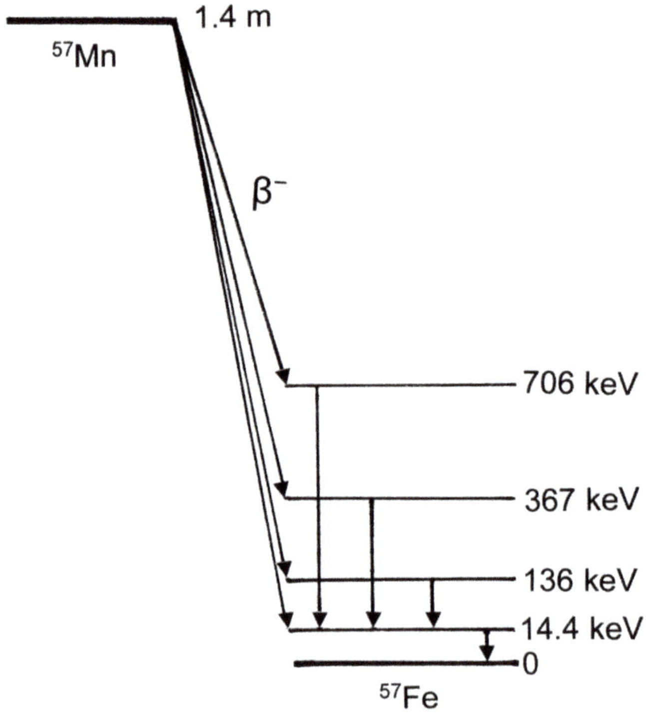

Figure 11.4. Simplified energy level diagram showing the population of the ^{57}Fe Mössbauer state from a 1.4 min half-life ^{57}Mn precursor. Principal decay modes are shown. Compare with figure 2.8.

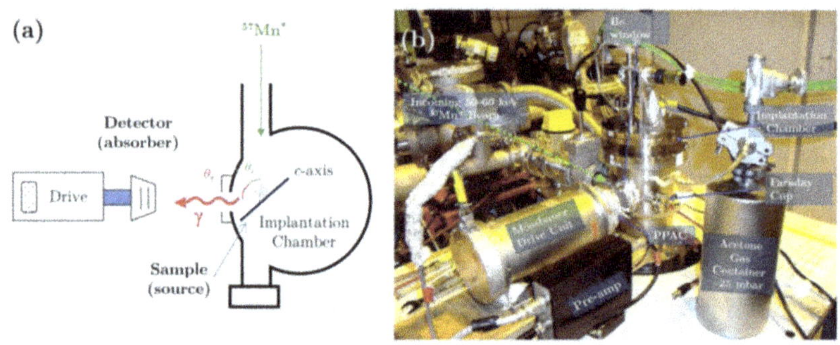

Figure 11.5. (a) Simplified schematic diagram of the ion implantation system and Mössbauer spectrometer used at ISOLDE for on-line $^{57}Mn/^{57}Fe$ Mössbauer spectroscopy. (b) Photograph of the apparatus. Figure 1 from Masenda *et al* (2022). Copyright by the authors (2022). CC BY 4.0.

the implantation system and Mössbauer spectrometer at the ISOLDE facility is shown in figure 11.5. Radioactive ^{57}Mn is produced by 1.4 GeV proton-induced fission in uranium carbide (UC_2). Separation of ^{57}Mn from other fission products has been described by Fedoseyev *et al* (1997). As in the previous section, the sample, after ion implantation, is used as the Mössbauer source with a moving single-line

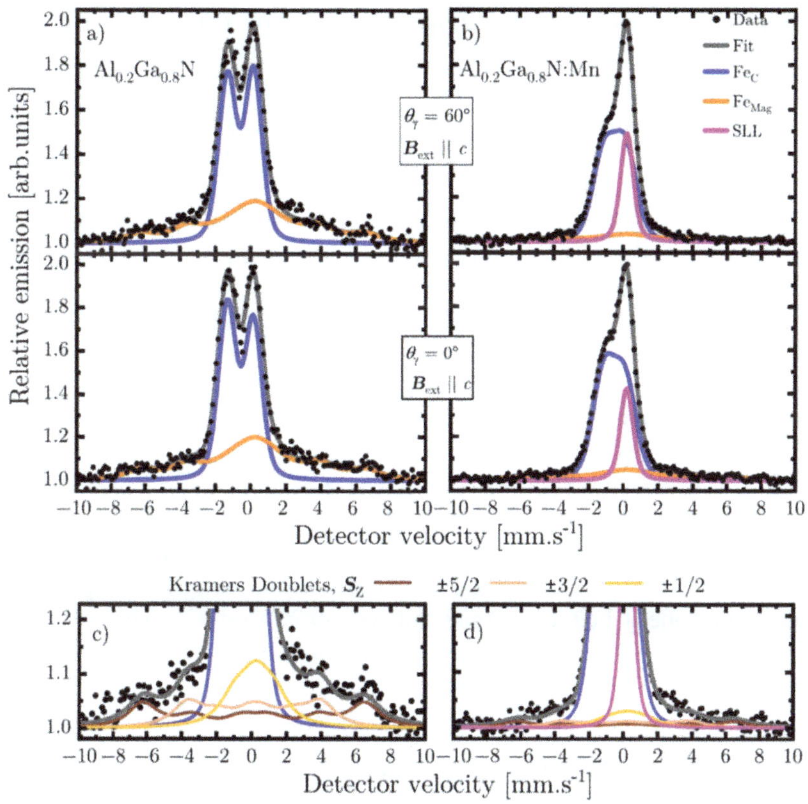

Figure 11.6. Room temperature ^{57}Fe Mössbauer effect spectra of ^{57}Mn implanted $Al_{0.2}Ga_{0.8}N$ (left) and $Al_{0.2}Ga_{0.8}N$:Mn (right) in an external magnetic field of 0.6 T parallel to the crystallographic *c*-axis. Results are shown for two sample orientations, $\theta_\gamma = 60°$ and $\theta_\gamma = 0°$, as indicated in the figure. Details of the fits to the spectra are discussed in the text. Note that the velocity scale on the horizontal axis of the figure represents the detector velocity and is inverted relative to that of a conventional moving source experiment. The velocity scale is calibrated relative to α-Fe. Figure 2 from Masenda H *et al* (2022). Copyright by the authors (2022). CC BY 4.0.

absorber. The angle between the sample *c*-axis and the γ-ray direction, θ_γ in the figure, can be varied for the Mössbauer measurements.

Masenda *et al* (2022) have undertaken studies of $Al_xGa_{1-x}N$ thin films prepared by metal-organic vapor phase epitaxy with and without 1 at% Mn doping, designated $Al_xGa_{1-x}N$ and $Al_xGa_{1-x}N$:Mn, respectively. Figure 11.6 shows some results from this study for $x = 0.2$. These results provide considerable information about the location and valence states of the ^{57}Fe probe atoms in the ion implanted structure and the effects of Mn doping.

In the undoped sample (figure 11.6(a)), the spectra consist of a central doublet with broadened wings on the sides. The wings can be observed in more detail in the enlarged spectra at the bottom of the figure. This component accounts for 43% of the spectral area and has an isomer shift of +0.12 mm s^{-1}. It corresponds to high-spin Fe^{3+} on substitutional Ga/Al sites. This spectral component was fitted to three magnetically split sextets corresponding to S_z states of ±5/2, ±3/2 and ±1/2 with a maximum hyperfine field of 40 T.

The central doublet accounts for the remaining 57% of the spectral area and was fitted to a Voigt-based doublet where the spectral asymmetry was modelled by isomer shift-quadrupole splitting correlation (see section 8.6). This component shows a mean isomer shift of +0.58 mm s^{-1} and a mean quadrupole splitting of 1.48 mm s^{-1}. This spectral component has been attributed to high-spin Fe^{2+} atoms located substitutionally on Ga/Al sites with neighboring nitrogen vacancies caused by the ion implantation.

The spectra of the Mn doped sample are illustrated in figure 11.6(b). These show the magnetically split component and quadrupole split components, as in the undoped sample, with an additional singlet component. The relative area of the Fe^{2+} doublet (63%) remains similar to that in the undoped spectra. The isomer shift (+0.49 mm s^{-1}) is similar to the undoped sample, but the quadrupole splitting (1.25 mm s^{-1}) is somewhat reduced, due to the presence of substitutional Mn atoms on the Ga/Al sites. The hyperfine parameters of the magnetic sites remain similar to those of the undoped sample, but the intensity of this spectral component has been reduced to about 12% at the expense of the additional singlet component at 24%. The singlet component has an isomer shift of −0.22 mm s^{-1}. This negative isomer shift is attributed to Fe^{4+} atoms on substitutional Ga/Al sites where the 4+ valence state is stabilized as a result of substitutional Mn^{2+} neighbors.

11.5 In-beam Mössbauer spectroscopy

The final category of ion implantation experiments utilizes a source nuclide with a half-life that is comparable to the half-life of the excited Mössbauer state. Several approaches to this technique have been utilized. For example, at Stanford University, the tandem Van de Graaff accelerator has been used to accelerate ^{35}Cl ions to 64 MeV. These ions were incident on an ^{57}Fe enriched foil which ejected excited ^{57}Fe ions that were then implanted into the sample under investigation (Latshaw *et al* 1980). At RIKEN a somewhat different approach has been taken (Kobayashi *et al* 2008). Naturally occurring (91%) ^{56}Fe nuclei in the sample of interest were thermal neutron irradiated to produce excited state ^{57}Fe by the reaction $^{56}Fe(n, \gamma)^{57}Fe$.

Here we look at some results from the Hahn-Meitner-Institut Berlin (Menningen *et al* 1987) which takes an approach similar to that at Stanford University. A schematic diagram of this in-beam Mössbauer experiment is shown in figure 11.7. A pulsed 130 MeV beam of ^{40}Ar ions was incident on a target iron foil enriched to 90% in ^{57}Fe. Excited state ^{57}Fe ions are ejected from the foil and are implanted into the sample of interest (catcher). A single-line ^{57}Fe enriched absorber and parallel-plate counter are attached to the Mössbauer drive unit to record the ^{57}Fe Mössbauer spectra. The sample is cooled in a liquid helium flow cryostat, meaning that the implantation and measurement are made at the same temperature.

Figure 11.8 shows some results from a study of Fe implanted aluminum by Menningen *et al* (1987) obtained at different implant/measurement temperatures.

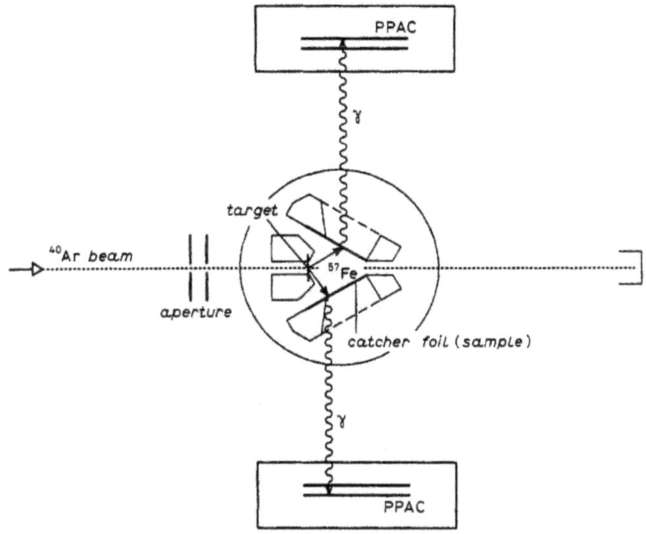

Figure 11.7. Diagram of the in-beam ^{57}Fe Mössbauer setup at the Hahn-Meitner-Institut Berlin. PPAC: parallel-plate avalanche counter. From Menningen *et al* (1987) copyright IOP Publishing. Reproduced with permission.

The spectra are appropriately fitted to a singlet (on the left side of the spectra in the figure) and an unresolved doublet (on the right side of the spectra in the figure). Following the results of other studies of Fe impurities in Al (see e.g., Dunlap *et al* 1988, Mansel *et al* 1973, Nasu *et al* 1974, Petry *et al* 1982, Verbiest and Pattyn 1982) the left-hand component is attributed to ^{57}Fe probe atoms that have entered into the fcc Al structure substitutionally. The measured isomer shifts for this component of the spectra, as shown in figure 11.8, are consistent with Fe impurity sites in Al, as discussed in section 9.6. The doublet is attributed to Fe which enters into the fcc Al structure interstitially.

The relative intensities of the two components at low temperature (5.5 K), are 40% substitutional and 60% interstitial. It has been reported (Mansel *et al* 1978) that the recoil-free fractions of the these two sites is the same at low temperature. Thus, the spectrum at 5.5 K is a direct measurement of the relative populations of the two sites. The spectra in figure 11.8 show a clear decrease in the relative area of the interstitial doublet component with increasing temperature. More specifically, as the temperature is increased there is a slight decrease in the area of the substitutional singlet component and a significant decrease in the area of the interstitial doublet. In absolute terms, the slight decrease in the area of the singlet is consistent with the expected temperature dependence of the recoil-free fraction as discussed in chapter 2. This means that the actual population of substitutional sites is temperature independent. The decrease in the area of the interstitial doublet is interpreted as anomalous changes in the temperature dependence of the recoil-free fraction for Fe atoms in interstitial sites. This effect is consistent with previous studies and is characterized by a drastic reduction in the recoil-free fraction above about 14 K

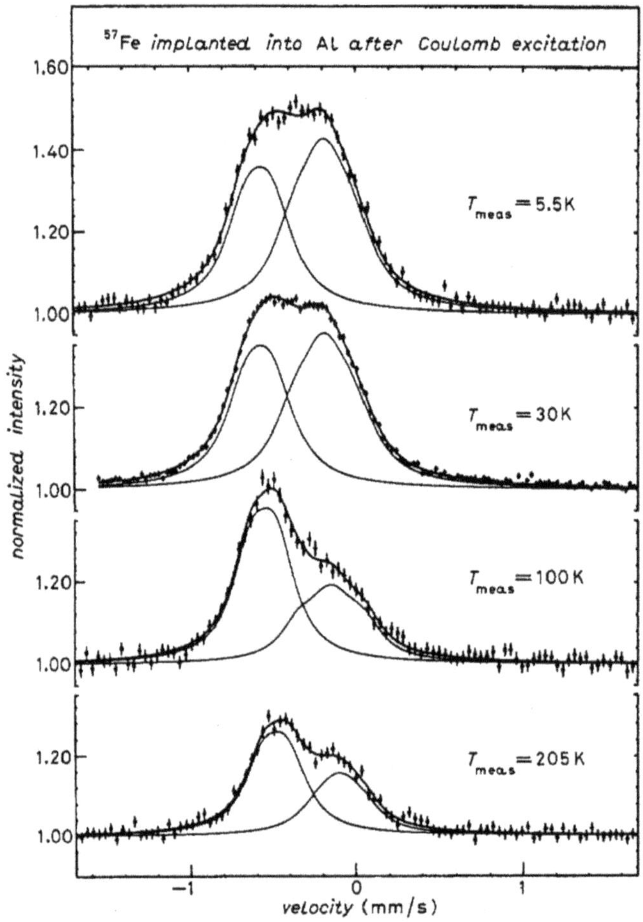

Figure 11.8. ^{57}Fe Mössbauer effect spectra for Fe implanted Al as a function of temperature for an in-beam experiment as described in the text. See note concerning velocity scale in the caption for figure 11.6. From Menningen *et al* (1987) copyright IOP Publishing. Reproduced with permission. All rights reserved.

which results from the onset of atomic jump processes (see Mansel *et al* 1973, Vogl *et al* 1976).

11.6 Fundamentals of synchrotron radiation

Mössbauer effect studies of materials that have low concentrations of an active nuclide or that require very rapid data accumulation, can be limited by the strength of conventional Mössbauer sources. As well, traditional Mössbauer spectroscopy is not possible when a suitable precursor for the Mössbauer transition is not available. While implantation studies, as discussed above, have offered one possibility to deal with such issues, synchrotron radiation-based Mössbauer techniques offer another. The very intense photon flux from synchrotron radiation sources, in particular, makes such an approach attractive. In order to describe the use of synchrotron

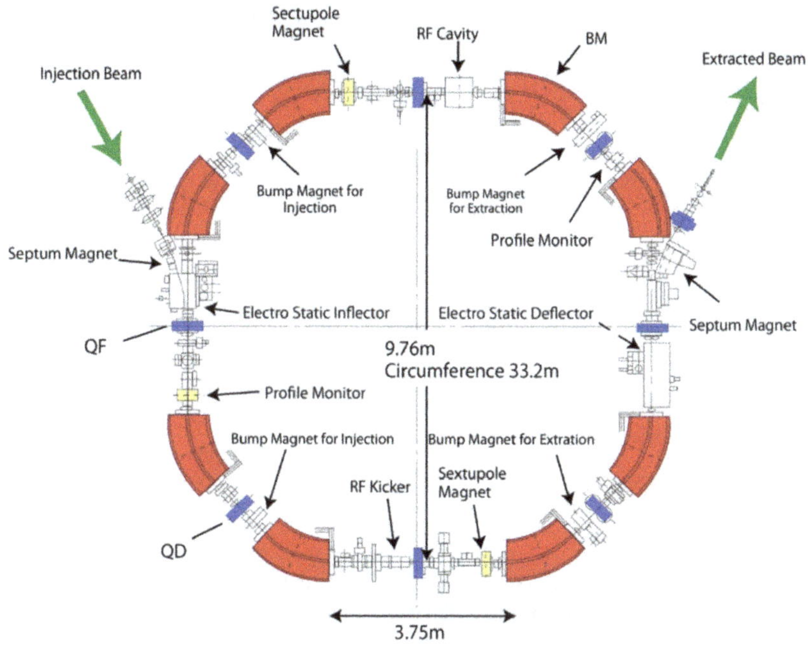

Figure 11.9. Schematic diagram of a synchrotron. Reprinted from Shiokawa *et al* (2021). Copyright (2021). With permission from Elsevier.

radiation for Mössbauer spectroscopy, we first provide a brief overview of the design and operation of a synchrotron.

The simplest cyclic particle accelerator, the cyclotron, uses a constant magnetic field and a periodic electric potential with a constant frequency to accelerate charged particles. For an overview of particle accelerators see Wiedemann (2015). This constant field, constant frequency approach is suitable as long as the particles are non-relativistic. At higher energies other techniques must be utilized. The synchrotron, as illustrated in figure 11.9, is one that uses both variable magnetic fields and variable accelerating frequencies. In this approach, the particle orbits can be constrained to be circular rather than spiraling outward, as in the cyclotron, and this minimizes the area over which the magnetic field needs to be applied.

Two of the important features of the synchrotron that are illustrated in figure 11.9 are the bending magnets (BM) and the accelerating (RF) cavity. As the particle energy is increased by successive passes through the accelerating cavity, the magnetic field strength of the bending magnets and the accelerating cavity frequency are increased so that the beam path remains within the magnetic field regions and the accelerating potential remains synchronized with the movement of particle bunches. For relativistic particles the applied magnetic field, B, is given by

$$B = \frac{\gamma m v}{q r}$$

where m is the particle rest mass, c is the speed of light, q is the particle charge, r is the radius of curvature of its path. γ is the Lorentz factor defined as

$$\gamma = \left[1 - \frac{v^2}{c^2}\right]^{-1/2} = \frac{E}{mc^2}$$

where E is the particle total energy. Under these conditions the frequency of the accelerating potential, f, is

$$f = \frac{qB}{2\pi\gamma m}.$$

An important characteristic of charged particles in a cyclic accelerator is the fact that they radiate. The power radiated in the form of photons is referred to as synchrotron radiation and the power, P, is given as,

$$P = \frac{1}{6\pi\varepsilon_0}\frac{q^2E^4}{c^7m^4r^2} \tag{11.1}$$

where ε_0 is the vacuum permittivity and r is the radius of curvature of the orbit. The radiation is emitted in the form of a forward cone, as illustrated in figure 11.10.

An important factor in equation (11.1) is the mass dependence of the synchrotron radiation power. The equation indicates that lighter particles, such as electrons, produce substantially more synchrotron radiation than heavier particles, such as protons or ions. For example, at the same energy electrons will radiate about $(2000)^4 \approx 10^{13}$ times as much power as protons. This energy must be replaced in order to maintain the particle energy, and this becomes increasingly more significant as the energy increases. For this reason, synchrotrons have been most useful in particle physics for the study of the properties of heavy particles. However, synchrotron radiation itself has found many applications in the study of materials. For example,

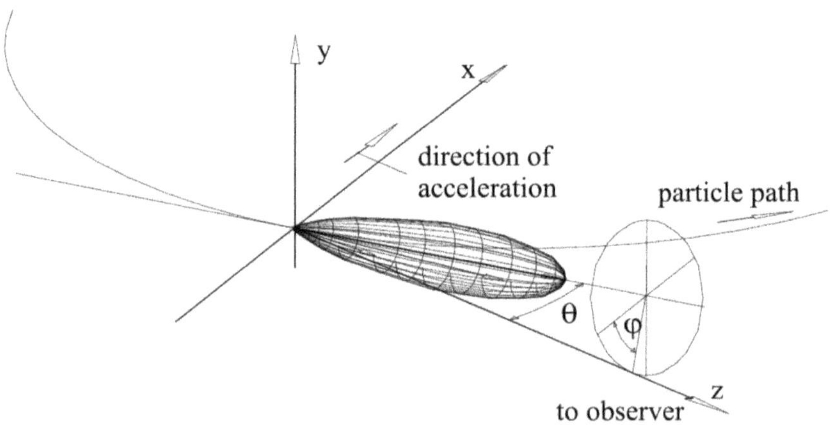

Figure 11.10. Geometry of synchrotron radiation. CC BY 4.0 From Wiedemann (2015) With permission of Springer.

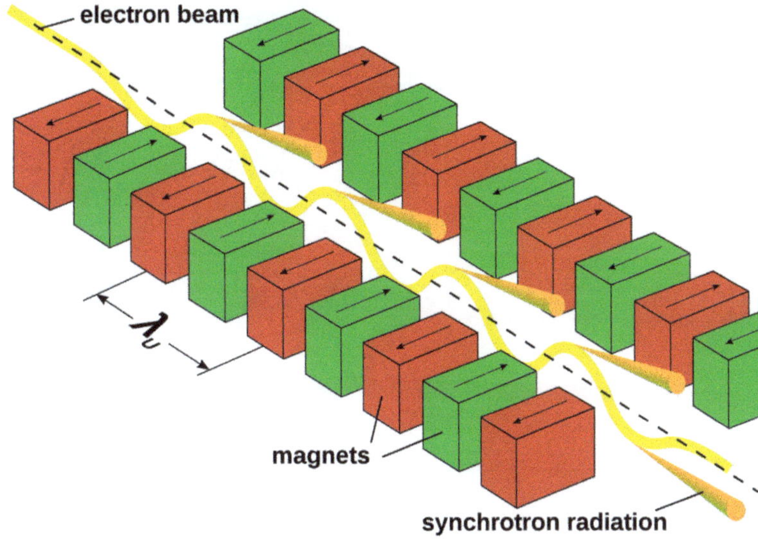

Figure 11.11. Design of an undulator for producing synchrotron radiation. This [undulator] image has been obtained by the author from the Wikimedia website where it was made available by Holst (2005) under a CC BY-SA 3.0 licence. It is included within this chapter on that basis. It is attributed to Bastian Holst.

the very intense x-rays that are produced have been utilized for x-ray diffraction and x-ray imaging experiments. Synchrotrons that are dedicated to the production of synchrotron radiation accelerate electrons to increase the intensity of this radiation. In order to optimize the intensity of synchrotron radiation the radius of curvature of the particle trajectory as indicated in equation (11.1), i.e., $1/r^2$, may be decreased by the inclusion of additional magnets in the beam path. These magnets are wiggler or undulator magnets and an example of an undulator magnet is shown in figure 11.11. Wiggler or undulator magnets are placed in the beam path of the synchrotron at locations where experiments are positioned.

To understand how synchrotron radiation can be used experimentally, it is important to look at the energy spectrum of the emitted photons. Figure 11.12 shows typical synchrotron radiation spectra as a function of photon energy for different electron beam energies. In all cases it is clear that the spectrum is very broadband and the peak energy increases with increasing beam energy. For electron beam energies of a few GeV and higher, the synchrotron radiation spectrum extends from a fraction of a keV to around 100 keV. In the next section we describe how very intense broadband synchrotron radiation can be used to observe Mössbauer spectra.

11.7 Synchrotron radiation-based Mössbauer techniques

As noted above, synchrotron radiation covers a wide range of energies in the x-ray portion of the electromagnetic spectrum. The intensity of this radiation makes it an attractive option for a variety of experiments that use traditional sources of radiation in the energy region. Studies where sample size, active nuclide

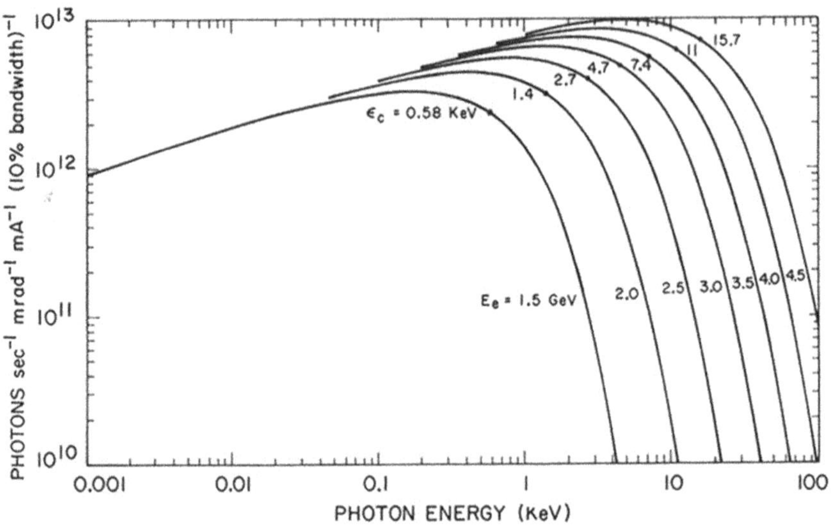

Figure 11.12. Energy spectrum of synchrotron radiation as a function of electron beam energy. This [syncrotron radiation emission as a function of the beam energy] image has been obtained by the author from the Wikimedia website where it was made available by Jackson (1999) under a CC BY 2.5 licence. It is included within this chapter on that basis. It is attributed to JD Jackson.

concentration or dynamic behavior are limiting factors for traditional experiments have benefited from the availability of synchrotron radiation. These applications fall roughly into three categories: diffraction, spectroscopy and imaging. Diffraction experiments typically probe microstructural properties, while spectroscopic studies typically probe composition and electronic structure. The imaging capabilities of synchrotron radiation have greatly extended the scope of medical and biological investigations.

Since x-rays in the 10 s of keV range are capable of exciting low-lying nuclear levels, synchrotron radiation has also found applications in various aspects of nuclear spectroscopy. Mössbauer spectroscopy is an ideal application of synchrotron radiation as the available range of photon energies, i.e., up to about 100 keV, covers the range of energies over which recoil-free fractions are compatible with viable Mössbauer studies. The application of synchrotron radiation to the Mössbauer effect has been recently reviewed by Alp *et al* (2012), Rüffer and Chumakov (2000) and Seto *et al* (2009).

In order to understand the means by which intense synchrotron radiation can be used to produce Mössbauer spectra we have to consider two important characteristics of this radiation. Firstly, it is broadband, as illustrated in figure 11.12 and, secondly it is pulsed. This latter feature results from the fact that the electrons travel around the synchrotron in bunches and produce pulses of synchrotron radiation with a duration of ~100 ps.

The basic principle of synchrotron radiation-based Mössbauer effect spectroscopy is to use the intense synchrotron radiation to excite the appropriate

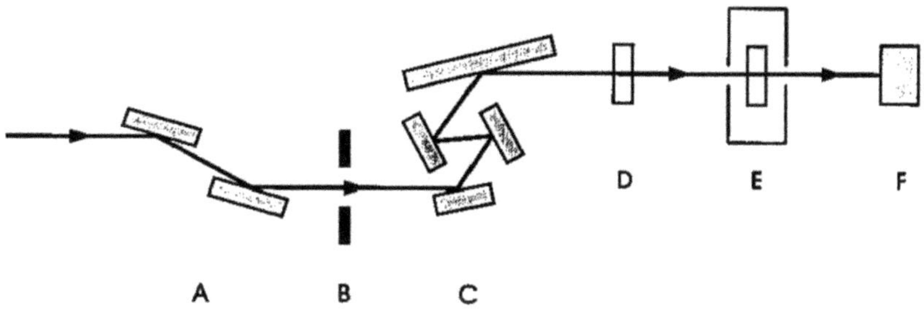

Figure 11.13. Typical two-stage Si crystal monochromator used to produce narrow bandwidth x-ray radiation. A, high-heat-load monochromator; B, slits; C, high-resolution, nested monochromator; D, iron foil; E, electron detector: F, photodiode. Reprinted with permission from Sturhahn *et al* (1996). Copyright 1996 by the American Physical Society.

Mössbauer transition in the nucleus of interest. This excited nucleus then serves as a source of Mössbauer γ-rays. The nucleus can be located in the material of interest, or it may be in reference material that produces a single-line γ-transition. In the latter case, the so-called synchrotron Mössbauer source takes the place of the single-line source in a conventional Mössbauer experiment.

To effectively excite only the appropriate Mössbauer state, the bandwidth of the incident radiation must be substantially reduced. This is done using a monochromator. Typically, a two-stage Si crystal monochromator as illustrated in figure 11.13, is used. The first stage consists of a high-heat-load monochromator which reduces the energy bandwidth to about 2 eV for the 14.4 keV peak of ^{57}Fe (for example). The second stage consists of a nested high-resolution monochromator which reduces the energy bandwidth to about 5 meV at the 14.4 keV peak of iron. The x-ray radiation produced by the monochromator is suitable for the excitation of the target nuclear level to populate the Mössbauer state and this can then be used as a radiation source for Mössbauer experiments.

Two basic approaches to using synchrotron radiation from the Mössbauer source as described above have been taken, time domain synchrotron Mössbauer spectroscopy and energy domain synchrotron Mössbauer spectroscopy. The former utilizes the pulsed nature of the synchrotron source, and the latter is more similar to conventional Mössbauer spectroscopy. These two approaches are described below.

11.8 Time domain synchrotron Mössbauer spectroscopy

The basic experimental configuration for time domain synchrotron Mössbauer spectroscopy is shown in figure 11.14. After passing through the monochromator, the x-ray beam is incident on the absorber, which is material under investigation. The absorber contains Mössbauer active nuclei compatible with the x-ray energy determined by the monochromator. It is important for the operation of the time domain synchrotron Mössbauer spectrometer that the x-ray beam is pulse. As noted above, the duration of the pulse is of the order of 100 ps. This is substantially shorter than the lifetime of any viable Mössbauer states, which are in the ns to tens of ns

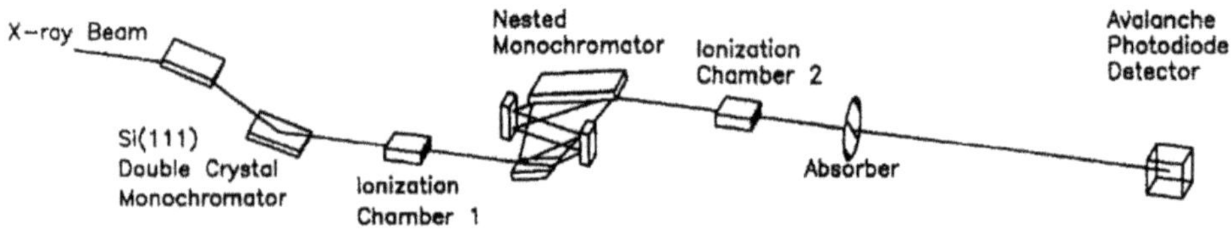

Figure 11.14. Diagram of spectrometer for observing time domain synchrotron Mössbauer effect spectra. Reprinted from Alp *et al* (1995). Copyright (1995). With permission from Elsevier.

range (see chapter 2). It is also important that the time between successive x-ray pulses is substantially longer than the Mössbauer state lifetime.

Thus, when the x-ray pulse is incident on the sample, the Mössbauer level will be populated and will begin to decay as given by the decay law in equation (2.5). As discussed in chapter 2, this exponential decay represents the behavior of the system in the time domain, while the behavior of the system in the energy domain is represented by the Lorentzian given in equation (2.6).

In a time domain synchrotron radiation Mössbauer experiment, a γ-ray detector, as shown in figure 11.14, measures the emission from the sample as a function of time after the x-ray pulse populates the Mössbauer level. In the case of emission from a single-line source, the time domain spectrum will merely show an exponential decay. In the case where the energy domain spectrum shows more than a single Lorentzian line, the time domain spectrum will develop beats. The beats result from two or more exponential decays that differ slightly in their decay constant and these can be fitted to an appropriate model to extract the Lorentzian pattern in the corresponding energy domain representation.

Lin *et al* (2011) have reported a time domain ^{57}Fe synchrotron Mössbauer investigation of the pressure dependence of Fe_2O_3. It is known that Fe_2O_3 undergoes a structural transition from a corundum-type phase to the Rh_2O_3(II)-type phase at 40–50 GPa (Pasternak *et al* 1999). It is also known that Fe_2O_3 undergoes a Mott insulator–metal transition as a function of pressure and Lin *et al* (2011) have looked at the relationship between these two transitions. They have used a ^{57}Fe enriched (95%) Fe_2O_3 sample as the absorber for a time domain synchrotron Mössbauer experiment and results are illustrated in figure 11.15. Time domain spectra exhibiting the characteristic beat pattern are shown in the left panel of figure 11.15(a) as a function of applied pressure. Measured data are shown by the black data points and the red lines show the fitted curves. From the fitted parameters, energy domain Mössbauer effect spectra can be reconstructed, as shown in the right panel of figure 11.15(a). Figure 11.15(b) shows the pressure dependence of the Mössbauer parameters, as obtained from the data in figure 11.15(a). The magnetic fraction clearly shows a decrease starting at a pressure around 54 GPa, corresponding to the structural transition. This structural transition corresponds to the onset of the insulator–metal transition which is complete by around 80 GPa, as evidenced by changes in the magnetic field and quadrupole splitting.

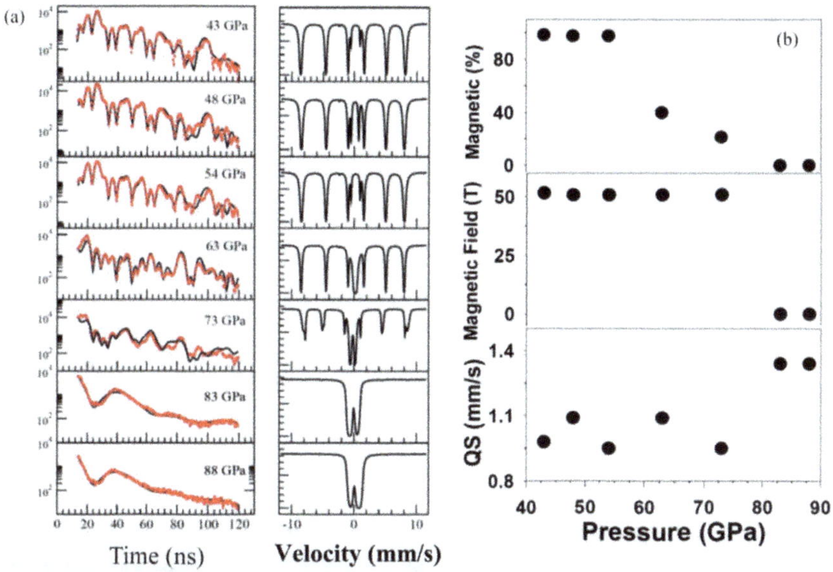

Figure 11.15. (a) Time domain ^{57}Fe synchrotron Mössbauer effect spectrum of $^{57}Fe_2O_3$ (left panel) and corresponding calculated energy spectra (right panel) for different applied pressures. (b) Pressure dependence of measured Mössbauer effect parameters. Reprinted with permission from Lin *et al* (2011). Copyright 2011 by the American Physical Society.

11.9 Energy domain synchrotron Mössbauer spectroscopy

Energy domain synchrotron Mössbauer spectroscopy uses either a single-line source, which is produced by synchrotron irradiation, with the material of interest as the absorber/scatterer, analogous to traditional Mössbauer absorber experiments, or it uses the material of interest as the source and a single-line absorber/scatterer analogous to traditional source (or emission) experiments and off-line ion implantation experiments. In the former case, the source is often referred to as a synchrotron Mössbauer source. Experiments may be conducted in either a transmission or scattering geometry. This approach produces a traditional Mössbauer effect spectrum as a function of velocity and allows for a conventional approach to data analysis.

Figure 11.16 shows a typical energy domain synchrotron Mössbauer experiment which uses a single-line $^{57}FeBO_3$ scatterer as a synchrotron Mössbauer source. The material under investigation (the absorber, A), is attached to a traditional Mössbauer velocity transducer. Figure 11.17 shows a somewhat different design using a scattering geometry. Either the transmitter or the scatterer can be the single-line reference, while the other is the sample under investigation.

Seto *et al* (2009) have reported ^{57}Fe energy domain synchrotron Mössbauer spectroscopy results that demonstrate the utilization of this technique. They have utilized a system with the sample under investigation as a stationary transmitter and

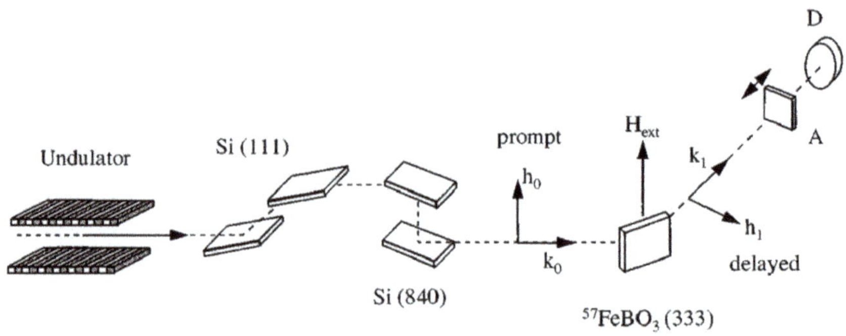

Figure 11.16. Diagram of spectrometer for observing energy domain synchrotron Mössbauer effect spectra in the transmission geometry. A = absorber, D = detector. Reprinted with permission from Smirnov *et al* (1997). Copyright 1997 by the American Physical Society.

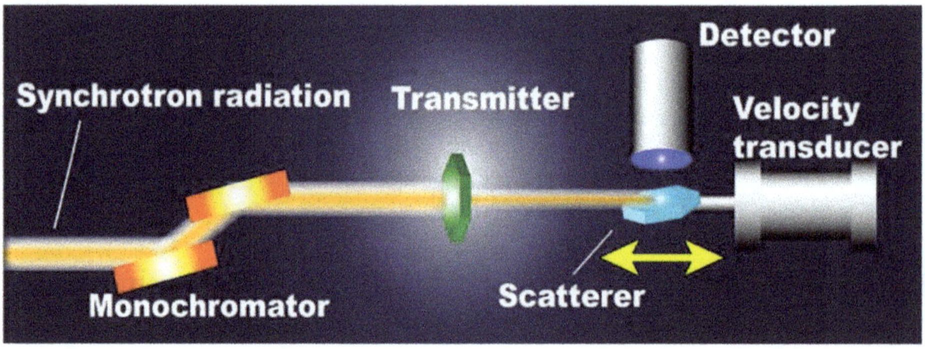

Figure 11.17. Diagram of spectrometer for observing energy domain synchrotron Mössbauer effect spectra in the scattering geometry. From Seto *et al* (2010) copyright IOP Publishing. Reproduced with permission. All rights reserved.

a reference sample as a moving scatterer, as illustrated in figure 11.17. Figure 11.18 shows a spectrum obtained using powdered α-Fe_2O_3 as the transmitter and a Pd foil containing 2 at% ^{57}Fe as the single-line scatterer. The spectrum was fit to a Lorentzian sextet and yielded hyperfine parameters consistent with conventional Mössbauer studies (e.g., van der Woude 1966).

Germanium is an element that can potentially be investigated by means of the Mössbauer effect and represents an excellent example of the benefits of synchrotron radiation-based techniques. The Mössbauer state of 7.7 % naturally abundant ^{73}Ge is the $7/2^+$ third excited state at 68.75 keV. This state has a half-life of 1.74 ns and decays primarily to the $9/2^+$ ^{73}Ge ground state (Singh 2004). The observation of the ^{73}Ge Mössbauer effect is very difficult (at best) using traditional techniques because of the very low yield of the Mössbauer state from the β^- decay precursor ^{73}Ga.

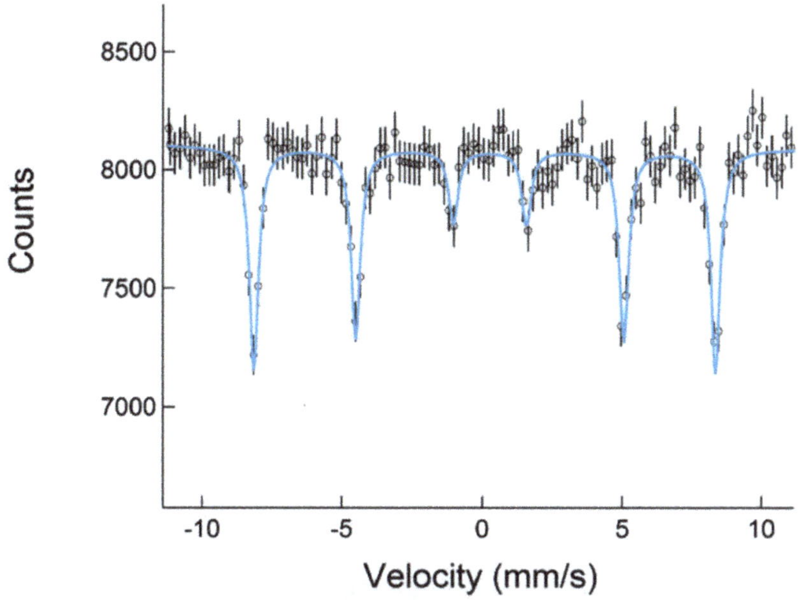

Figure 11.18. Room temperature energy domain synchrotron Mössbauer spectrum of $\alpha\text{-}Fe_2O_3$ from Seto *et al* (2009) obtained as described in the text. Reprinted with permission from Seto *et al* (2009). Copyright 2009 by the American Physical Society.

Successful measurements have been made using the implantation of Coulomb excited ^{73}Ge nuclei into a host material (Czjzek *et al* 1968) along the lines of the approach described in section 11.5. Synchrotron radiation methods are, therefore, an attractive means of observing ^{73}Ge Mössbauer spectra.

Seto *et al* (2009) have reported the results of a ^{73}Ge synchrotron Mössbauer experiment. These measurements used a powdered Li_2GeO_3 transmitter as the sample under investigation and a powdered tetragonal GeO_2 scatterer as the reference sample. In the tetragonal GeO_2 structure ($P4_2/mnm$) each Ge ion is surrounded by six octahedral oxygen. Although the octahedral environment is slightly distorted, no ^{73}Ge quadrupole splitting has been observed (Czjzek *et al* 1968), and GeO_2 serves as a suitable single-line reference.

Figure 11.19 shows the ^{73}Ge Mössbauer spectrum of Li_2GeO_3 obtained by Seto *et al* (2009) as described above. The spectrum exhibits a broadened singlet with an isomer shift of -0.21 mm s^{-1} relative to the GeO_2 reference sample. There is no indication of quadrupole splitting and, as Li_2GeO_3 is non-magnetic, it is assumed that there is no hyperfine magnetic splitting. The observed broadening is likely due to thickness effects (see section 8.2). This example shows the importance of synchrotron radiation-based Mössbauer spectroscopy for the investigation of viable Mössbauer transitions when there is no suitable radioactive precursor.

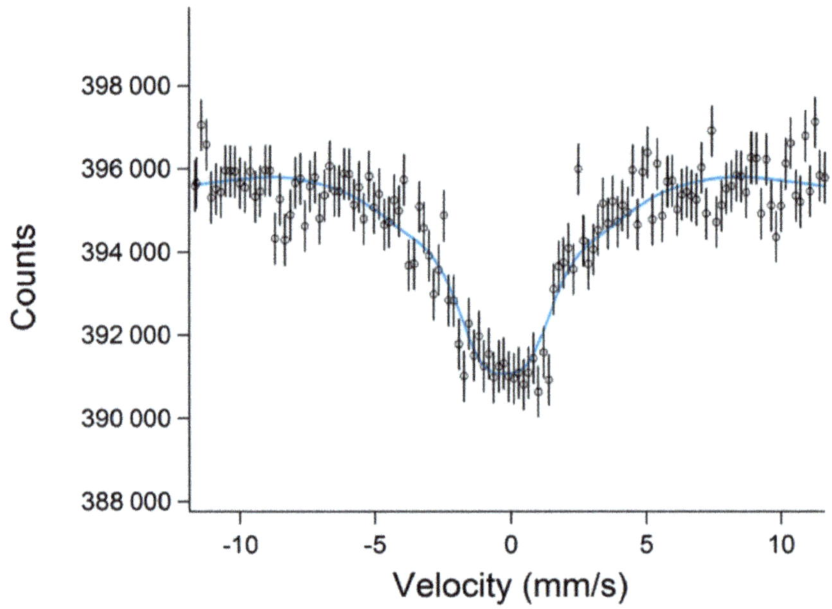

Figure 11.19. ^{73}Ge synchrotron radiation Mössbauer spectrum of Li_2GeO_3 referenced to GeO_2 as described in the text. Reprinted with permission from Seto *et al* (2009). Copyright 2009 by the American Physical Society.

Bibliography and suggestions for further reading

Alp E E, Sturhahn W and Toellner T 1995 Synchrotron Mössbauer spectroscopy of powder samples *Nucl. Instrum. Methods Phys. Res.* **97** 526–9

Alp E E, Sturhahn W, Toellner T S, Zhao J and Leu B M 2012 Nuclear resonance scattering of synchrotron radiation as a unique electronic, structural and thermodynamic probe *The Rudolf Mössbauer Story—His Scientific Work and Its Impact on Science and History* ed M Kalvius and P Kienle (Berlin: Springer) pp 339–56

Armini A J, Bunker S N and Spitzer M B 1982 Non-mass-analyzed ion implantation equipment for high volume solar cell production *Proc. 16th IEEE Photovoltaic Specialists Conf.* pp 895–9

Czjzek G, Ford J L C, Love J C, Obenshain F E and Wegener H H F 1968 Coulomb-recoil-implantation Mössbauer experiments with ^{73}Ge *Phys. Rev.* **174** 331–45

Dunlap R A 2023 *An Introduction to the Physics of Nuclei and Particles* 2nd edn (Bristol: IOP Publishing)

Dunlap R A, Lloyd D J, Christie I A, Stroink G and Stadnik Z M 1988 Physical properties of rapidly quenched Al-Fe alloys *J. Phys. F: Met. Phys.* **18** 1329–41

Fedoseyev V N *et al* 1997 Chemically selective laser ion source of manganese *Nucl. Instrum. Methods Phys. Res.* B **126** 88–91

Gaudet J M, Hatchard T D, Farrell S P and Dunlap R A 2008 Properties of Fe–Ga based powders prepared by mechanical alloying *J. Magn. Magn. Mater.* **320** 821–9

Hastings J B, Siddons D P, van Bürck U, Hollatz R and Bergmann U 1991 Mössbauer spectroscopy using synchrotron radiation *Phys. Rev. Lett.* **66** 770–3

Holst B 2005 *Undulator* (https://commons.wikimedia.org/wiki/File:Undulator_(english).svg)

Jackson J D 1999 *Syncrotron Radiation Emission as a Function of the Beam Energy* (https://commons.wikimedia.org/wiki/File:Syncrotron_radiation_emission_as_a_function_of_the_beam_energy.png)

Jha S, Suyanto D, Julian G M, Dunlap R A, Cheong S W, Fisk Z and Thompson J D 1990 Mössbauer studies of ^{57}Co doped layered perovskites *Hyperfine Interact.* **55** 1323–6

Jin Q Y, Liu Y G, Zhou Y, Wu Q, Zhai Y J and Sun L T 2021 RF and microwave ion sources study at Institute of Modern Physics *Plasma* **4** 332–44

Kitao S, Masuda R, Fujihara T, Tajima H, Nagasawa N, Yoda Y, Masuda T, Yoshimura K and Seto M 2022 Development of time- and energy-resolved synchrotron-radiation-based Mössbauer spectroscopy *J. Phys. Conf. Ser.* **2380** 012136

Kobayashi Y *et al* 2008 Neutron in-beam Mössbauer spectroscopy of iron disulfide at 298 and 78 K *Hyperfine Interact.* **187** 49–55

Kobayashi Y, Yoshida Y, Yoshida A, Watanabe Y, Hayakawa K, Yukihira K, Shimura F and Ambe F 2000 In-beam Mössbauer study of ^{57}Mn/^{57}Fe in Si following projectile fragmentation and implantation *Hyperfine Interact.* **126** 417–20

Langouche G 1992 The Mössbauer search for Fe in Si *Hyperfine Interact.* **72** 215–28

Langouche G and De Potter M 1987 Identification of substitutional and interstitial Co implanted in Si *Nucl. Instrum. Methods Phys. Res.* B **19–20** 322–4

Latshaw G L, Russell P B and Hanna S S 1980 Ion-beam implantation of iron into germanium, silicon and diamond studied by the Mössbauer effect *Hyperfine Interact.* **8** 105–27

Langouche G and Yoshida Y 2013 Ion implantation *Mössbauer Spectroscopy: Tutorial Book* ed Y Yoshida and G Langouche (Heidelberg: Springer) ch 6 pp 267–301

Leupold O, Pollmann J, Gerdau E, Rüter H D, Faigel G, Tegze M, Bortel G, Rüffer R, Chumakov A I and Baron A Q R 1996 Nuclear resonance scattering of synchrotron radiation at the 21.5 keV resonance of ^{151}Eu *Europhys. Lett.* **35** 671–6

Lin J F, Tse J S, Alp E E, Zhao J, Lerche M, Sturhahn W, Xiao Y and Chow P 2011 Phonon density of states of Fe_2O_3 across high-pressure structural and electronic transitions *Phys. Rev.* B **84** 064424

Mansel W, Vogl G and Koch W 1973 Direct evidence for interstitial-atom trapping by Co^{57} impurities in aluminum from Mössbauer-effect measurements after low-temperature neutron irradiation *Phys. Rev. Lett.* **31** 359–62

Mansel W, Meyer H and Vogl G 1978 Interstitial atom clustering at ^{57}Co impurity atoms in aluminium after electron and neutron irradiation *Radiat. Eff.* **35** 69–77

Masenda H *et al* 2022 Unusual charge states and lattice sites of Fe in $Al_xGa_{1-x}N$:Mn *New J. Phys.* **24** 103007

Menningen M, Sielemann R, Vogl G, Yoshida Y, Bonde-Nielsen K and Weyer G 1987 Interstitial implantation of iron into aluminium 1987 *Europhys. Lett.* **3** 927–33

Nagy D L 1994 Trends in Mössbauer emission spectroscopy of ^{57}Co/^{57}Fe *Hyperfine Interact.* **83** 1–19

Nasu S, Gonser U, Shingu P H and Marakami Y 1974 ^{57}Fe Mössbauer spectra in splat quenched Al-0.5, 1, 3 and 5 at% Fe alloys *J. Phys. F: Metal Phys.* **4** L24–8

Pasternak *et al* 1999 Breakdown of the Mott–Hubbard state in Fe_2O_3: a first-order insulator–metal transition with collapse of magnetism at 50 GPa *Phys. Rev. Lett.* **82** 4663–6

Petry W, Vogl G and Mansel W 1982 Mössbauer study of localized diffusion in an interstitial cage *Z. Phys. B—Condens. Matter* **46** 319–29

Rüffer R 2008 Nuclear resonance scattering *C. R. Phys.* **9** 595–607

Rüffer R and Chumakov A I 2000 Nuclear inelastic scattering *Hyperfine Interact.* **128** 255–72

Sage J T, Paxson C, Wyllie G R A, Sturhahn W, Durbin S M, Champion P M, Alp E E and Scheidt W R 2001 Nuclear resonance vibrational spectroscopy of a protein active-site mimic *J. Phys. Condens. Matter* **13** 7707–22

Sawicka B D and Sawicki J A 1977 Evidence of the electric quadrupole coupling of ^{57}Fe implanted in silicon *Phys. Lett.* A **64** 311–2

Schwen D 2005 *Schematics of an Mass Separating Implantation/Deposition Setup* (https://commons.wikimedia.org/wiki/File:Ion_implanter_schematic.png)

Seto M, Masuda R, Higashitaniguchi S, Kitao S, Kobayashi Y, Inaba C, Mitsui T and Yoda Y 2009 Synchrotron-radiation-based Mössbauer spectroscopy *Phys. Rev. Lett.* **102** 217602

Seto M, Masuda R, Higashitaniguchi S, Kitao S, Kobayashi Y, Inaba C, Mitsui T and Yoda Y 2010 Mössbauer spectroscopy in the energy domain using synchrotron radiation *J. Phys.: Conf. Ser.* **217** 012002

Seto M, Yoda Y, Kikuta S, Zhang X W and Ando M 1995 Observation of nuclear resonant scattering accompanied by phonon excitation using synchrotron radiation *Phys. Rev. Lett.* **74** 3828–31

Shiokawa T, Okugawa Y, Kurita T and Nakanishi T 2021 Slow beam extraction method from synchrotron for uniform spill and fast beam switching using an RF knockout method of multi-band colored noise signal—POP experiment and simulation *Nucl. Instrum. Methods Phys. Res.* **1010** 165560

Singh B 2004 Nuclear data sheets for A=73 *Nucl. Data Sheets* **101** 193–323

Smirnov G V, van Bürck U, Chumakov A I, Baron A Q R and Rüffer R 1997 Synchrotron Mössbauer source *Phys. Rev.* B **55** 5811–5

Sturhahn W *et al* 1995 Phonon density of states measured by inelastic nuclear resonant scattering *Phys. Rev. Lett.* **74** 3832–5

Sturhahn W, Quast K W, Toellner T S, Alp E E, Metge J and Gerdau E 1996 Electron emission from ^{57}Fe nuclei excited with synchrotron radiation *Phys. Rev. B: Condens. Matter* **53** 171–5

Toellner T S, Alp E E, Graber T, Henning R W, Shastri S D, Shenoy G and Sturhahn W 2011 Synchrotron Mössbauer spectroscopy using high-speed shutters *J. Synchrotron Radiat.* **18** 183–8

van der Woude F 1966 Mössbauer effect in α-Fe_2O_3 *Phys. Status Solidi* B **17** 417–32

Verbiest E and Pattyn H 1982 Study of radiation damage in aluminum *Phys. Rev.* B **25** 5097–121

Vogl G, Mansel W and Dederichs P H 1976 Unusual dynamical properties of self-interstitials trapped at Co impurities in Al *Phys. Rev. Lett.* **36** 1497–500

Wiedemann H 2015 *Particle Accelerator Physics* (Cham: Springer)

Weyer G 2000 Mössbauer spectroscopy at ISOLDE *Hyperfine Interact.* **129** 371–90

Xie X, Chen C, Luo J and Xu J 2021 Effect of nitrogen ion implantation energy on the mechanical and chemical properties of AISI M50 steel *Int. J. Chem. Eng* **2021** 4630661

www.ingramcontent.com/pod-product-compliance
Ingram Content Group UK Ltd.
Pitfield, Milton Keynes, MK11 3LW, UK
UKHW051853100926
8556IPUK00005B/610

9 780750 360371